AF735161

UNDERSTANDING AFRICAN REAL ESTATE MARKETS

This book brings together a broad range of research that interrogates how real estate market analysis, finance, planning, and investment for residential and commercial developments across the African continent are undertaken. In the past two decades, African real estate markets have rapidly matured, creating the conditions for new investment opportunities which has increased the demand for a deeper understanding of the commercial and residential markets across the continent. The chapters consider issues that pertain to formal real estate markets and the critical relationship between formal and informal property markets on the continent.

With contributing authors from South Africa, Ghana, Nigeria, Uganda, Kenya, and Tanzania, the book considers the achievements of African real estate markets while also highlighting the complex central themes such as underdeveloped land tenure arrangements, the availability of finance in both the commercial and residential sectors, rapidly growing urban areas, and inadequate professional skills. This book is essential reading for students in real estate, land management, planning, finance, development, and economics programs who need to understand the nuances of markets in the African context. Investors and policy makers will learn a lot reading this book too.

Aly Karam is a visiting associate professor in the School of Architecture and Planning at University of the Witwatersrand, Johannesburg, South Africa. He earned a PhD in city and regional planning from the University of North Carolina at Chapel Hill, North Carolina. His research interests and publications extend to both the planning and the architecture disciplines. In architecture he publishes on marketing and on the use of IT in practice and its effect on the management of the architecture offices. His planning research and publications revolve around informality, land and real estate, housing policy, and the housing process.

François Viruly is an associate professor and director of the Urban Real Estate Research Unit at the University of Cape Town, South Africa. He has been President and Executive Director of the African Real Estate Society and the 2022 President of the International Real Estate Society. His research interest lies in the development and sustainability of property markets across the African continent.

Catherine Kariuki is a lecturer at the Department of Real Estate, University of Nairobi. She holds a Bachelor of Arts degree in land economics and a Master of Arts degree in housing administration and is a registered and licensed valuer. Her research interests are in property management, valuation, housing and land management and administration. She is a past President of the African Real Estate Society and a board member of the International Real Estate Society. She has previously served as a board member of the Valuer's Registration Board.

Victor A. Akujuru holds a BSc degree in valuation and estate management from Bristol Polytechnic, a MSc in urban land appraisal from the University of Reading, and a PhD in built environment from Salford University, all in England. He is a fellow of the Nigerian Institution of Estate Surveyors and Valuers, a registered estate surveyor and valuer, a Right of Way Agent, and a Right of Way Instructor. He has held visiting research positions at Kaduna State University, and University of Jos, Nigeria. He is a past president of the African Real Estate Society.

International Real Estate Markets

The Chinese Real Estate Market
Development, Regulation and Investment
Junjian Cao

Real Estate in Italy
Markets, Investment Vehicles and Performance
Guido Abate and Guiditta Losa

Real Estate Markets and Development in South America
Understanding Local Regulations and Investment Methods in a Highly Urbanised Continent
Claudia Murray, Eliane Monetti and Camilla Ween

Real Estate in South Asia
Edited by Prashant Das, Ramya Aroul and Julia Freybote

Real Estate in Central America, Mexico and the Caribbean
Claudia Murray, Camilla Ween, Yadira Torres and Yazmin Ramirez

The Real Estate Market in Ghana
An Emerging Market in Sub-Saharan Africa
Wilfred K. Anim-Odame

UNDERSTANDING AFRICAN REAL ESTATE MARKETS

Edited by Aly Karam, François Viruly, Catherine Kariuki and Victor A. Akujuru

Routledge
Taylor & Francis Group
LONDON AND NEW YORK

Cover image: © Getty Images

First published 2022
by Routledge
4 Park Square, Milton Park, Abingdon, Oxon OX14 4RN

and by Routledge
605 Third Avenue, New York, NY 10158

Routledge is an imprint of the Taylor & Francis Group, an informa business

© 2022 selection and editorial matter, Aly Karam, François Viruly, Catherine Kariuki and Victor A. Akujuru; individual chapters, the contributors

The right of Aly Karam, François Viruly, Catherine Kariuki and Victor A. Akujuru to be identified as the authors of the editorial material, and of the authors for their individual chapters, has been asserted in accordance with sections 77 and 78 of the Copyright, Designs and Patents Act 1988.

All rights reserved. No part of this book may be reprinted or reproduced or utilised in any form or by any electronic, mechanical, or other means, now known or hereafter invented, including photocopying and recording, or in any information storage or retrieval system, without permission in writing from the publishers.

Trademark notice: Product or corporate names may be trademarks or registered trademarks, and are used only for identification and explanation without intent to infringe.

British Library Cataloguing-in-Publication Data
A catalogue record for this book is available from the British Library

Library of Congress Cataloging-in-Publication Data
A catalog record has been requested for this book

ISBN: 978-0-367-23308-2 (hbk)
ISBN: 978-0-367-23310-5 (pbk)
ISBN: 978-0-429-27925-6 (ebk)

DOI: 10.1201/9780429279256

Typeset in Bembo
by Apex CoVantage, LLC

CONTENTS

List of Contributors *vii*
Acknowledgements *xi*

1 Sub-Saharan Real Estate Markets: An Overview of the Contributions 1
François Viruly and Aly Karam

2 Global trends and African real estate markets 4
Wilfred K. Anim-Odame

3 African Cities, Urban Growth Management, and Property Development 18
Cecil Madell and Aly Karam

4 The residential, office, retail and industrial markets in West Africa 37
Emmanuel Kofi Gavu, Anthony Owusu-Ansah and Wilfred K. Anim-Odame

5 Land Tenure Systems and Sustainable Land Management With Special Reference to the Kenyan Rural Areas 48
Jennifer Murigu

6 The impact of land tenure on real estate markets and land-use rights and land registration 63
Terzungwe T. Dugeri and Austin C. Otegbulu

7 Towards a Harmonious Land Administration Institutional Service Delivery in Kenya; a Decentralized Approach 75
Winnie Mwangi

8 The Policy Frameworks Supporting Sustainable Development of Real Estate in Kenya 91
Mwenda K. Makathimo

9 An Analysis of the Evolution of the Sub-Saharan African Residential Real Estate Research 105
Aly Karam, François Viruly, and Taiwo Afinowi

10 Inclusive, Affordable, and Smart Housing in Africa 122
Prisca Simbanegavi and Kolawole Ijasan

11 A tale of two pathways: How the low-end Gauteng residential property market may be shaped either by state-driven supply or by a more enabled private sector 135
Napier Mark, Cooper Antony K, Holloway Jennifer P, Thanjekwayo Londeka, Graham, Nick and le Roux Alize

12 Appraisal of the Nigerian real estate market: Constraints and opportunities 158
Austin C. Otegbulu and Victor A. Akujuru

13 Challenges of Raising Real Estate Finance in Nigeria 168
Akinola. E. Olawore and Austin C. Otegbulu

14 Performance of Sector-Specific and Diversified REITs in South Africa 176
Omokolade Akinsomi

15 Application of Building Management System (BMS) in Commercial Property Management in Kenya 188
Walter Ogolla and Raphael Kieti

16 An Investigation Into the Perceived Impact of Implementing Innovative Digital Technology in the Property Transaction Process in South Africa: Implications for African Property Markets 201
Jason Green, François Viruly and Alireza Moghayedi

17 Real Estate Education in Africa South of the Sahara 213
Catherine Kariuki, Kathy Michell and Rachael Daisy Mirembe

18 The Real Estate Education in West Africa 228
Victor A. Akujuru and Christopher L. Deeyah

Index *251*

CONTRIBUTORS

Taiwo Afinowi is an architect, urbanist, and doctoral candidate at the Witwatersrand School of Architecture and Planning. He holds a master's of built environment in housing and a professional master's in architecture. His doctorate explores the intercessions between global urban agendas in housing and practice at local levels. His research interests include housing, urban renewal and regeneration, urban agendas and governance, and global policy implementation.

Akinola. E. Olawore is Principal Partner and Group Chairman of Akin Olawore & Co and its subsidiaries. He is also a director and partner of Blacksands Mosets, a maritime security company. He was a partner of International Realty Valuers Inc., San Diego, USA, and was also appointed Adjunct Professor of Real Estate from 2012 to 2014 by the Homburg Academy, Prince Edward Island, Canada. He was a member of a number of professional associations, including: a fellow of the Nigerian Institution of Estate Surveyors and Valuers, Nigeria; fellow, Royal Institution of Chartered Surveyors, UK; fellow, Certified Auctioneers Institute of Nigeria, Nigeria; registered surveyor and valuer; fellow, Institute of Management Consultants of Nigeria; member, Institute of Chartered Arbitrators, UK; member, Institute of Rwandan Property Valuers, Rwanda; professional member, Association for Project Management, UK; full member, Standing Conference of Mediation Advocates, UK; professional member, Nigerian Institute of Management (Chartered); RICS accredited civil and commercial mediator; and a member of the RICS President's Panel on Commercial ADR.

Omokolade Akinsomi holds a PhD in real estate finance and investments from National University of Singapore. He has held visiting research positions at the National University of Singapore, University of Tokyo, Japan, and the University of Western Sydney, Australia. He is the Vice-President of the African Real Estate Society.

Wilfred K. Anim-Odame holds a PhD in finance, with specialisation in real estate finance from Cass Business School, City University, London. He was the Chief Executive Officer of the Ghana Lands Commission and currently is a senior technical advisor at the National Development Planning Commission, Ghana. He has published extensively, is a fellow of the Ghana Institution of Surveyors, and is a member of the Royal Institution of Chartered Surveyors.

Antony K. Cooper is a principal researcher in Smart Places at the CSIR in Pretoria, South Africa. He is also extraordinary lecturer, Department of Geography, Geoinformatics and Meteorology, University of Pretoria. His recent work has covered smart cities, geoinformatics, and ethics. He is an Extraordinary Lecturer at the University of Pretoria and a registered professional GISc practitioner.

Christopher L. Deeyah is a lecturer in the Department of Estate Management, Rivers State University (RSU), Port Harcourt, Nigeria. He is a PhD candidate in environmental valuation. His research interest includes environmental valuation, land management, and sustainable development. He is an associate member of the Nigerian Institution of Estate Surveyors and Valuers.

Terzungwe T. Dugeri studied estate management at University of Nigeria for BSc, and University of Lagos for MSc and PhD. Currently Reader in the Department of Estate Management, University of Jos, he researches property markets and involuntary resettlement. He is a senior partner at Ter Dugeri, Anyam & Partners.

Emmanuel Kofi Gavu (Dr.-Ing.) is a lecturer at the Department of Land Economy, Kwame Nkrumah University of Science and Technology (KNUST) Kumasi, Ghana. His main fields of teaching and research are: application of GIS in urban management, real estate, and housing markets analysis. Emmanuel is a board member of the African Real Estate Society (AfRES), Vice Chair of the Future Leaders of the African Real Estate Society (FLAfRES), a member of Ghana Institution of Surveyors (GhIS), and a member of the Society of Property Researchers (giF) Germany.

Nick Graham is an urban geographer and registered professional engineer with master's degrees in civil engineering, environmental policy, and urban geography. For the past 12 years, he has undertaken public policy research with a focus on long-term infrastructure planning and municipal financial modelling for local and national government initiatives.

Jason Green holds a BSc (Honors) property studies degree from the University of Cape Town (UCT), as well as a Bcom finance degree from the Independent Institute of Education. Jason is a research assistant at UCT and is employed as an investment manager at a real estate focused family office. He has interests in property technology, finance, and investments.

Jennifer P. Holloway is a senior statistician at CSIR with an MSc in mathematical statistics from the University of Cape Town. Her expertise lie in research and application of quantitative solutions to complex problems, particularly in the area of multivariate statistical modelling. She is a chartered member of ICCSSA and a member of SASA.

Kolawole Ijasan, Associate Professor at the School of Construction Economics and Management, is a housing market regeneration expert, having worked on numerous projects in South Africa, Nigeria and the UK. His research interests range from real estate valuation to urban economics. He is a NRF rated researcher.

Raphael Kieti holds a bachelor of arts degree in land economics, a master of arts in valuation and property management, and a doctorate of philosophy in land economics, all from the University of Nairobi in Kenya. He has research interests in property valuation, building maintenance management, facilities management and Building Automation System. Dr Kieti is currently a lecturer in the Department of Real Estate at the University of Nairobi.

Mwenda K. Makathimo is the Executive Director of Land Development and Governance Institute. He has wide experience in practice and research on issues relating to land and environment policy development, land management and administration, governance, assets and property valuation, real estate management, boundary delimitation, and mapping. He is a registered and licensed valuer, holds a bachelor of arts degree in land economics, a master of arts in valuation and property management, and a doctorate in environmental policy from the University of Nairobi. He has also undertaken executive training on extractive industries and sustainable development at the Columbia University. He is a past chairman of the Institution of Surveyors of Kenya and was also a commissioner of the Interim Independent Boundaries Review Commission. He has served as a member of the Nairobi City Valuation Court, an external examiner at the University of Nairobi, School of the Built Environment, Department of Real Estate, and Senior Lecturer Mount Kenya University, School of Engineering, Energy and the Built Environment.

Kathy Michell holds a PhD in facilities management from the University of Salford. Kathy's core teaching passions are in the fields of property management and facilities management. Her research areas are in the application of urban and community-based facilities management in leveraging socio-economic development in African cities.

Rachael Daisy Mirembe is a property economist with over 14 years' experience in the property industry. She is a lecturer in Makerere University Business School on the real estate programme. Rachael has published many articles with international real estate journals. She is a member of ERES, PRRES, AFRES, and AREA (U). Her research interests are in real estate markets, real estate education, PropTech, and sustainability. She is currently pursuing a PhD (property studies) at the University of Cape Town (SA).

Alireza Moghayedi is senior research officer, founder, and co-director of Sustainability oriented + Cyber Unit for Built Environment (S+CUBE) at the University of Cape Town. He holds a PhD in construction economics and management, MS in civil engineering major in transportation, MS in construction management, and BS in civil engineering.

Jennifer Murigu is a senior lecturer in the University of Nairobi's Department of Real Estate. A land economist by profession, her main areas of interest are in property and facilities management and land management. Jennifer, who is the lead author of an award winning paper in the 10th Afres Conference, lives in Nairobi.

Winnie Mwangi, a senior lecturer in the Department of Real Estate at University of Nairobi, holds a BA in land economics (University of Nairobi), MSc in housing in development (UCL – London) and a PhD in land economics (University of Nairobi). Winnie's research interest is in institutional governance in Land Administration.

Mark Napier is a principal researcher at the CSIR in Pretoria and visiting professor at the University of the Free State. Mark graduated as an architect from the University of KwaZulu-Natal and studied housing at postgraduate level at Newcastle University, UK. His research interests include urban knowledge management, land and housing markets, housing transformations, home-based enterprises, and many aspects of informal settlements.

Walter Ogolla holds a bachelor's degree in real estate from the Technical University of Kenya and a master of arts in valuation and property management from the University of Nairobi in Kenya. He has research interests in property valuation, facilities management, and Building Automation System. He is currently a graduate assistant in the Department of Real Estate and Property Management at the Technical University of Kenya.

Austin C. Otegbulu obtained a BSc degree in estate management UNN, MSc in construction management UNILAG, and PhD in environmental management with focus on environmental valuation. He lectures at the University of Lagos (between 1988 and 1999 as an associate lecturer and from 1999 to date on full time). He also is the Founding Partner of A. C. Otegbulu & Partners, a firm of estate surveyors and valuers; past National Secretary of Nigerian Institution of Estate Surveyors; and current member of its Board of Trustees.

Anthony Owusu-Ansah is an associate professor and Head of the Accounting and Finance Department at the Business School of the Ghana Institute of Management and Public Administration (GIMPA), Accra. He is the author of the book *Construction and Application of Property Price Indices* published by Routledge. He has published extensively in the areas of house price index construction, housing market dynamics, hedonic modelling, and rental housing market mechanisms. He is a professional member of the Ghana Institution of Surveyors and an associate of the British Higher Education Academy.

Alize le Roux is a geoinformatics researcher with extensive experience in providing planning and decision support solutions to local, regional, and national decision-makers. She holds a master's degree in geographical sciences from the University of Utrecht, Netherlands, specialising in multi-hazard risk assessments and spatial decision support systems. She is currently a senior researcher at the Institute for Security Studies, South Africa.

Prisca Simbanegavi is a senior lecturer at Wits University, South Africa. She holds a BSc (Honors) in economics, a Bcom (Honors) in finance, an MSc in real estate, and a PhD in real estate. Her research interests are in inclusive residential developments/investments. She is a member of AfRES and GWIRE.

Londeka Thanjekwayo holds a master's in the built environment from the University of the Witwatersrand, Johannesburg, and a postgraduate diploma in developing social housing projects from Erasmus University, Rotterdam. Her interests lie in exploring pragmatic solutions through research and policy. She headed the research in the Gauteng Department of Human Settlements and currently is at the City of Johannesburg as the National Programme Liaison & Policy (IGR).

ACKNOWLEDGEMENTS

This book is the work of many colleagues and friends from across the African continent. Their contributions, time and effort over a long period have played a considerable role in the publication of this book. The original thought of the book grew over several years which included discussions during the African Real Estate Society annual conferences. While attending the conferences we realized the wealth of knowledge and research that is conducted on the continent across a number of sectors and in numerous countries. Unfortunately, a considerable amount of research does not see the light of day beyond the annual conference and even fewer make it to the *Journal of African Real Estate Research*. Another encouragement included the realization that our teaching, professional and research community lacked an academic book that consolidated our present understanding of African real estate markets as well as the critical differences that exist between countries of sub-Saharan Africa.

The book includes areas of research that have received considerable attention in numerous conferences, seminars and international educational across the continent.

By compiling this publication comprising contributions of academics and professionals across the African continent we have hopefully displayed the vibrancy African property markets. Whilst these property markets have their specific characteristics, which drive local research agendas, growing attention is being given to global issues such as the affordability of housing markets, the relationship between residential property markets and social wellbeing, and the implications of sustainability agendas on property markets. The rapidly growing maturity of African property markets is also illustrated by the number of contributions that we received that deal with the growth of financial structures such as REITS valuation standards and ethics.

We trust that this publication will not only contribute in furthering the research output focused on African property markets, but that it will also encourage further research by African researchers. In addition, one of the critical objectives of this project was to provide educators and students with an educational text that will further the development of academics and professionals across the continent.

Finally, I would like to thank the dedicated team comprising François Viruly, Catherine Kariuki and Victor A. Akujuru for the perseverance in seeing this project to its conclusion. The project would also not have been possible without the encouragement and leadership of Ed Needle and Patrick Hetherington from Taylor & Francis; thank you for your continuous support and your unwavering belief that we will get to the end of this project.

We trust that our readers will find this book of interest and enjoyable.

1

SUB-SAHARAN REAL ESTATE MARKETS

An Overview of the Contributions

François Viruly and Aly Karam

Real estate markets across the African continent continue to receive considerable academic attention, reflecting the evolution that these markets have seen over the decades. The academic output suggests that residential and commercial markets in Africa have characteristics that need to adapt to new economic and social realities. While early research that deals with African property markets was largely focused on the residential property sector and land titling, the literature has evolved responding to a wider research agenda that now encapsulates investments and development of direct and indirect commercial property. In understanding African property markets it is important to appreciate that the maturity of property markets varies widely between countries across the continent. While countries such as South Africa and Mauritius are relatively transparent, other markets remain opaque and constrained by formal and informal institutional arrangements that have failed to keep up with market dynamics. In many countries across the African continent the banking sector remains underdeveloped and unable to meet the changing needs of property markets.

The contributions in this book can be classified under several themes. The first theme deals with the broad and often unique characteristics of African property markets. The second deals with legal considerations, placing a focus on the legal frameworks that drive property markets in Africa. The third looks at the dynamics of both residential and commercial property markets. The fourth considers the adoption of technologies in Africa, and the final theme looks at real estate education across the African continent.

The book starts with a contribution by Anim-Odame which considers the unique legal and financial arrangements that define African property markets. It also considers the impact of such arrangements on decision making and market outcomes. Anim-Odame concludes that in recent decades, global market forces have defined the trajectory of African markets as well as the improvements in market transparency in these markets. He also argues that a fully functioning and well-regulated real estate market can make a considerable contribution to national economic and social development.

Madell and Karam consider the role that property markets play in defining growth, urban form, and structures of African cities. The chapter emphasizes the considerable development problems that African cities grapple with, while also reminding us of the considerable level of

DOI: 10.1201/9780429279256-1

diversity existing between cities in Africa which in turn require unique interventions. They suggest that for city administrators the challenge lies in meeting the standards of a 'World Class City' while at the same time responding to an urban environment that often fails to meet the needs of the poor.

Gavu et al consider the institutional arrangements that characterize commercial property markets in West African property markets. The authors observe that governments can strengthen markets through the development of appropriate policies and stronger institutions, which result in increased market transparencies and greater domestic and international investment.

Several authors deal with land titling arrangements systems across the continent. Murigu deals with the evolution of legal land frameworks in East Africa, which range from leasehold to customary and freehold land tenure systems. She makes the critical point that in East Africa specific land tenure systems attempt to meet often unique socio-economic and sustainability objectives. The point is also made by Dugeri and Otegbulu who suggest that land tenure systems influence market outcomes and the potential to unleash value in property markets. They call for a strengthening of existing market arrangements in Nigeria. Mwangi places a focus on land administration systems and the potential impact that decentralization administration systems can have in meeting spatially defined socio-economic objectives.

Makathimo considers the relationship between land sustainability and legal arrangements in Kenya. The contribution considers the Kenyan constitutional provisions that require the state to respond to global sustainability objectives and agreements. The point is made that sustainability regulations have a considerable influence in the way that properties are developed, acquired, and disposed of.

Karam et al review the extensive literature that has dealt with the African residential property sector. They underline that the academic interest in African property markets has not only grown but has widened in scope and geography. The contribution identifies a literature that has, over the years, tackled numerous sub-themes such as affordability of housing, housing policy objectives, determinants of property values, housing finance, and the modelling of property markets. They also emphasize that the institutional complexities of African property markets retain a central position in the African real estate research agenda. They suggest that qualitative research remains constrained by a lack of data.

The dynamics of residential African property markets is also considered by Simbanegavi and Ijasan who posit that public sector interventions, such as inclusionary housing interventions, have a role to play in stimulating private sector investment through the reduction of risk. The duality of African property residential property markets is considered by Napier et al, who look at the dynamics that drive residential property markets. They argue that an accurate modelling of these would make it easier to identify market segments and fine tune public and private sector interventions.

Otegbulu turns to the commercial property sector, and in particular the impact that weak data, unreliable institutions, poor infrastructure, and an uncertain macro-economic environment have for property market outcomes. The commercial property sector is also considered by Olawore who emphasizes the importance financial markets play in developing markets in Africa. Akinsomi looks at the performance of REITS in South Africa and draws the conclusion that there is evidence that specialized REITs provide superior average returns.

The focus on technology in real estate is considered by several authors. Ogolla and Kiete turn to the technologies used in the management of properties and the specific role of

Building Management System (BMS). They conclude that demand is constrained by the low level of market awareness that exists for these types of technologies. They also assess the cost of applying technologies that are available to the market. The focus on appropriate technologies in property markets is also raised by Green et al who consider the role of PropTech in improving market transactions efficiencies in South Africa, as well as the opportunities and risk associated with such technologies.

The importance of education in promoting property markets is discussed by Kariuki et al who emphasize the development of property markets in Africa needs to be reflected in the development of appropriate curricula. Property courses should reflect the growing sophistication of markets. They suggest that the development of a body of knowledge needs to reflect existing market structures and complexities. The role of education in developing African property markets is also considered by Akujuru and Deeyah who highlight that modern property courses should place an appropriate emphasis on finance, valuation, and economics, while at the same time finding an appropriate mix of both 'technical 'and 'soft skills'. They underline that courses should reflect the interdisciplinary nature of the property body of knowledge.

The overarching theme of the contributions made in this book is that property markets in Africa are rapidly evolving and modernizing. These contributions also suggest that property market outcomes are influenced by the formal and informal institutions. These arrangements affect market efficiencies, market risk, and transaction costs which in turn influence investor confidence. As emphasized by several contributors, property markets in Africa need to meet diverse developmental objectives.

The challenge lies in developing an investment environment that can unleash economic as well as social value in African property markets.

2

GLOBAL TRENDS AND AFRICAN REAL ESTATE MARKETS

Wilfred K. Anim-Odame

1. Introduction

This chapter treats the African real estate markets under four broad segments and focuses on the impact of global trends. First, it examines the globalisation of real estate as an investment vehicle and highlights the limitations of African markets. It further covers the fundamentals of real estate investment based on available literature on selected African markets. Second, it assays the development of real estate markets in sub-Saharan Africa (SSA) to identify commonalities and uniqueness. Here there is a large literature on land administration and land policy but relatively little work on rigorous quantitative analysis of markets maturity, which is of direct relevance to the objectives of this chapter. Third, and in more detail, the chapter examines global trends under nine thematic segments – macroeconomic regimes, migration patterns, urbanisation, demographics, foreign direct investments, remittances, PropTech, mortgage market and covid 19 pandemic – to elucidate their impact on real estate markets in SSA. Fourth, it draws conclusions to include that a fully functioning and well-regulated real estate market largely depends on a number of fundamentals such as infrastructure development, information technology, the law, and policies that do not change with change of governments.

The past decades have experienced an influx in cross-border flows of real estate investments and an even more dramatic increase in the flow of real estate capital from advanced countries into emerging markets particularly in Asia and the transition economies of Central and Eastern Europe. The globalisation of real estate investment is accompanied by a growth in real estate market research in emerging and transition economies, from both academic and industry sources. Firms of real estate intermediaries have also expanded the coverage of the standard market indicators available in mature real estate markets to embrace a widening range of "emerging" markets in Eastern Europe, Asia and Latin America.

The increase in global real estate capital flows is a major factor for the improvement in transparency and information of a set of "emerging" real estate markets (Anim-Odame, 2009). In all these interventions, African real estate markets have been less affected. The business environment and size of higher-quality real estate investment stocks in African markets,

DOI: 10.1201/9780429279256-2

with the exception of South Africa, place them well outside the investment horizons of these investors. This narrative has to change while at the same time efforts are to be made to promote the real estate markets in SSA; make both domestic and foreign capital flows drive the process.

From a theoretical perspective and the limited evidence on actual values of the stocks, real estate plays a major role in the process of economic development. The lack of literature to explore this role is not only surprising but also suggests the need to investigate the operations of real estate markets in emerging economies. However, in developed economies, real estate is a large element of total capital stocks and of total personal and corporate wealth (Miles, 1990; Miles, Pittman *et al.*, 1991; Miles, Roberts *et al.*, 1994). At the global level, for example, Brown and Matysiak (2000) estimates that real estate – broadly defined as commercial and residential as well as farmland and other interests – accounts for 55 percent of the total value of wealth.

For developed countries, investible real estate value is set at the fixed ratio of 45 percent of GDP (Hughes and Arissen, 2005). For developing countries including emerging economies in sub-Saharan Africa, investible real estate value is rather a rising diminishing function of GDP per capita, tending towards the limit of 45 percent at the threshold of $20,000 per capita. Simplistic as it may be, the estimation method tends to exclude factors such as population density and planning regimes, which might plausibly be expected to lead to varying real estate values for given levels of GDP and GDP per capita. Nonetheless, the methodology quantifies the plausible propositions that, in terms of orders of magnitude, the value of real estate stocks is strongly associated with the size and wealth of economies.

Research and policy on the development of formal real estate markets in sub-Saharan Africa economies have concentrated on establishing land registration systems, on the assumption that improved security of tenure is the key element in the process. Other necessary conditions such as access to sources of equity and debt finance for development and investment, and the creation of clear rental contracts have not, however, been covered in the available literature. The attention of researchers and investors in income-producing real estate is heavily skewed towards developed economies, leaving emerging economies, particularly in Africa with small but rapidly growing formal markets uninformed by the standard market indicators such as rental growth, yields, total returns and market stock.

As demonstrated by a wealth of research in developed economies, a substantial benefit of the production of real estate indices is to inform investment markets of the case for the inclusion of real estate in mixed-asset portfolios. In emerging countries, and above all in African countries, information on the status of real estate as an investment medium – its market value against other asset classes, and its importance in the portfolios of institutional and private investors – is thin or non-existent. Similarly, there is very little literature on the availability of finance that was secured against real estate assets, or on the importance of real estate lending in the balance sheets of the banking system.

Data transparency is a key characteristic of a mature market, and markets that are able to demonstrate that they possess such data have a competitive advantage and are more likely to attract private sector investment funds (Adair *et al.*, 2004). The most systematic and extensive classification of real estate markets by indicators of data transparency and other measures of "maturity" can be found in a "Global Real Estate Transparency Index" (GRETI) produced by Jones Lang LaSalle. This is the real estate industry's most widely used and highly valued benchmark for assessing real estate transparency. The index covers a global list of countries that provide the most transparent operating environment for real estate investors, developers and corporate occupiers. In 2020, GRETI covered 99 countries with UK, US, France, Australia, Canada, New Zealand, Netherlands, Ireland, Sweden and Germany as the world's leading real estate investment destinations. These are real estate markets that are pushing the boundaries of transparency through technology, sustainability, anti-money laundering regulations and enhanced tracking of alternative sectors. These markets are also characterised by increasingly strong regulatory regimes, governance structures and efficient processes for real estate transactions.

In Africa, real estate market fundamentals such as access to credible data and the assessment of market performance remain patchy and must improve. Few countries have therefore been covered by GRETI. Out of 99 countries covered, Egypt (60), Morocco (61), Zambia (64), Nigeria (68), Rwanda (78) and Ghana (80) are the only economies in Africa that were covered but classified as low transparent with their respective rankings in brackets. As nascent real estate markets, they have tracking and regulation systems in the process of being developed. Other real estate markets including Angola (85), Mozambique (89), Uganda (90), Ivory Coast (92), Tanzania (95) and Ethiopia (98) have also been covered and classified as opaque. These are markets that are often hampered by geopolitical and economic challenges.

The majority of African countries lacks active real estate investment markets, and to a large extent, the primary sources of market evidence. But these are rather the drivers needed in the advanced economies and in a growing list of transitional and emerging economies to shape real estate research. Very few firms of global intermediaries such as Jones Lang LaSalle (JJL), Knight Frank, Broll and CBRE produce real estate market performance indicators for African countries.

Constructing real estate price and rental indices demands a substantial base of property investors, who will conduct open market valuations of their assets to international standards at least once a year. It is unlikely that this condition can currently be met in most African countries outside South Africa. Any efforts to create measures of market performance will probably require alternative methodologies, such as indices based on real estate transactions data (see Anim-Odame *et al.*, 2010).

Given the lack of primary data, published research on African real estate markets has consisted mainly of studies that are narrow in scope or focusing on the workings of informal markets with limited works on the nascent formal real estate markets in higher quality modern stock. Outside South Africa, published research on African real estate markets has predominantly focused on institutional structures, market's dynamics and constraints, and market's development.

Specifically, this chapter will serve to expand the knowledge base on global trends and their impacts on real estate markets in SSA through an exploration of selected themes namely macroeconomic regimes, migration patterns, urbanisation, demographics, foreign direct investments, remittances, property technology (PropTech), mortgage market and corona pandemic (covid 19).

2. Macroeconomic regimes

The global economy continues to be challenged with policy uncertainties, unstable growth, spikes in commodity prices and populist pressures in the run-up to elections in some countries, particularly in SSA. Strategic interventions are needed to address the escalating inflationary rates, fiscal imbalances and declining economic growth rates in most African economies. One of these is domestic revenue mobilisation that is needed to achieve debt sustainability and provide fiscal space for national development in general, and investment in infrastructural and human capital in particular.

Whilst economic growth continues to expand steadily in developed economies, the outlook for countries in Africa remains sluggish (IMF, 2018). Where modest economic recoveries have occurred, mainly due to oil exports as in Nigeria, and in particular for South Africa, average growth for sub-Saharan African economies increased from 2.7 percent in 2017 to 3.1 percent in 2018 (IMF, 2018).

Both macroeconomic and microeconomic factors are key determinants of real estate prices and rents. There is sufficient evidence in several studies on developed economies. Such major macroeconomic determinants of real estate pricing include GDP, inflation, interest rate and availability of funding such as mortgages (see Manganelli, 2014; Post and Berkhout, 2014; Ciarlone, 2015; Wang *et al.*, 2018; Lee, 2009). Microeconomic determinants of real estate prices and rents cover a set of characteristics of local environment where properties are located and other unique features of the subject property. For sub-Saharan African real estate markets, though limited, some works have provided evidence (Anim-Odame *et al.*, 2009, 2010; Owusu-Ansah, 2018). Surprisingly, previous studies on the impact of macroeconomic factors on real estate markets performance in SSA are virtually non-existent (Owusu-Ansah *et al.*, 2020).

A stable macroeconomic regime underpins real estate market development. Lending rates and Treasury bills rates are peaked in most economies in the sub-region. The resulting impact on real estate markets in SSA is two-fold: high cost of capital in the mortgage market and high-priced real estate assets. Much of these can be tackled through a reduction in domestic borrowing by governments. However, further works are extremely important to showcase the impact on the economies in the sub-region and also to inform policy decisions.

3. Migration patterns

The movement of people in and out of their territorial boundaries is termed migration. When the boundaries crossed are international, the movement of people into a country is referred to as immigration, whilst the movement out connotes emigration. The place of immigration is usually described as inbound country and that for emigration on the other hand is the outbound country. Humanity has been on the move since time immemorial. The search for work most often determines where people move to, and in some cases, investment potentials or economic opportunities were the driving force. Some migrants do so for many other reasons including family reunion, study, escape from conflict, human rights violation, persecution or in response to the adverse effect of natural disaster or other environmental effects.

In all these situations, three questions remain relevant to real estate markets:

i. Where are people moving to? In other words, which are the countries and cities people want to live in? Countries with better land use planning and land administration systems, favourable tax regimes, job opportunities, political stability, housing affordability, ease of doing business and investment opportunities tend to be attractive.

ii. What is going to happen in the inbound countries now, in the medium term and long term respectively? This is important because of housing needs, increase demand for services including pipe-borne water, electricity and telephone as well as pressure on infrastructure such as road networks as a result of the potential increase in vehicular and human traffic. The capacity of social services such as schools and hospitals will also be subjected to high demand as a result of the movement of people.
iii. Where should investors inject capital? Real estate markets tend to compete with all other sectors of the economy for investment. Competing investment vehicles will present various opportunities to prospective investors.

Real estate markets in Africa will require and also benefit from migration patterns as more people seek to live in one of its countries other than the one in which they were born. For example, in 2019, the number of migrants globally reached 272.0 million, an estimated 51.0 million more than the figure a decade previous (UN DESA, 2019). It is also estimated that 26 percent of the international migrants reside in America, 30 percent in Europe, 31 percent in Asia and 10 percent in Africa whilst only 3 percent settled in Oceania – Australia, New Zealand, Fiji, Papua New Guinea, among others. Of this figure, 3.5 percent constitutes international migrants, compared to 2.8 percent and 2.3 percent in 2000 and 1980 respectively.

The 2030 Agenda for Sustainable Development importantly recognises the role of migration. As such, 11 out of the 17 Sustainable Development Goals (SDGs), representing 65 percent, contain indicators and targets relevant to migration. This developmental agenda builds a good foundation for a better outlook for real estate markets, particularly in Africa where the concept of internal migration is deeply associated with urbanisation.

4. Urbanisation

The world's population for the next 10 years is projected to grow at an annual rate of 1.8 percent (Cohen, 2006). At this rate of growth, population can be expected to double globally by 2044. More than 50 percent of the world's population lives in cities and it is predicted that by 2030, around 61 percent of the population will be living in urban areas, at which time the world's urban population will be approaching 5 billion. Therefore, by 2050, the urban population will reach 6.5 billion. Cities will continue to serve as the main engine of economic growth and account for 75 percent of economic activities.

Global trends in urbanisation are re-shaping the world economic order with real estate markets having a key role to play in respect of demand for urban housing. But urbanisation has also accounted for a number of challenges such as slums, congestion, air pollution, social issues and pressure on real estate markets in most cities around the world. Population influx into cities makes housing the most basic demand with its associated inelastic supply. Housing in short supply invariably escalates prices in the market, and the reverse also holds, *ceteris paribus*.

Sub-Saharan Africa is becoming increasingly urban. Urbanisation is rapid with urban population growth averaged at 4.4 percent annually since the 1960s. With a total urban population of 450.7 million in 2019 and the increasing rate of urbanisation, the continent will continue to experience pull factors such as structural transformation, human capital accumulation and housing deficits in its major cities. Other consequences are overconcentration of growth and development in some cities; land use disorder and unplanned urban expansion; weak rural-urban linkages; inadequate urban investment and financing; and weak urban transportation planning and traffic management. Though increasing urbanisation is a global concern, its rapid rate that results in large-scale land use changes as in the case of SSA, is of course, a matter of concern.

Urbanization and city growth are caused by a number of factors including rural-urban migration, and natural population increase. In most countries in SSA, the pace of urbanisation has far outstripped planning capacity of city authorities. Several cities or towns have emerged without adequate spatial planning, which is not unexpected. It has unsurprisingly resulted in only 10 percent of urban land being devoted to roads; compared to 30 percent in cities in developed countries. It is extremely important to caution that poor spatial planning today stores up expensive problems for the future. Beyond sustainable spatial planning, cities require effective and efficient land administration, and systematic infrastructure investment. Arguably, this is necessary because when a city loses its boundaries, it ceases to be.

Unlike most East Asian economies, a number of sub-Saharan Africa countries are plagued with weak economies. The pace of increase in income per capita has not kept up with the pace of urbanisation. Certainly, these countries are becoming more urban but not simultaneously richer. The situation presents a serious mismatch for urban financing and planning. The consequences include congestion in the cities, high real estate pricing, housing deficit and high cost of transportation. Meanwhile, achieving the sustainable development goal 11 means ensuring access to safe and affordable housing, upgrading slums and improving public transport, providing green and public spaces, and improving urban spatial planning and management in all cities and towns.

Building key infrastructure will define the shape of cities in SSA, but governments cannot do it alone. Private sector participation is therefore urgently needed. City authorities can undoubtedly capture land value appreciation created by urbanisation through efficient and effective property taxation systems and the implementation of sustainable urban renewal projects. As Africa continues to urbanise at a rapid rate, policymakers must be proactive to enact policies that will stimulate cities' growth to achieve economic development and poverty reduction.

More rigorous studies on trends of real estate pricing and its consequences on economies in SSA may be extremely essential. These works would generate a new set of information and data that could not only promote real estate markets but also improve urban land planning and zoning. This effort is expected to contribute to solving many of the urban housing and environmental problems.

5. Demographics

It is estimated that worldwide, 26 percent of the population is under 15 years of age and 9 percent is over 65 years of age (Statista, 2020). In Africa, the gap is much wider, with 41 percent of the population being under 15 years old and only 3 percent being over 65 years old. Asian countries have 24 percent of their population being under 15 years old and 9 percent being over 65 years old. Countries in North America and Europe have swapped population figures; 18 percent of population in North America is under 15 years and 16 percent being over 65 years. As the figures stand, older population has outnumbered the young in many developed countries.

Countries in North America and Europe may be confronted with demographic challenges. For instance, as the ratio of young to old people falls, so societies are likely to see a falling working age population; government finances strained by higher spending on pensions, health and social services; and slower growth in tax revenues. These circumstances are expected to result in far-reaching effects on people's lifestyle, expenditure and working habits.

Sub-Saharan Africa is undergoing a major demographic transition and expected to yield some outcomes in the coming years. The total population of the sub-region has more than doubled over the past quarter century, increasing from 470 million in 1990 to 1.0 billion in 2019. This is projected to reach 1.7 billion by 2040 and close to 4.0 billion by 2099. The rest of the world is, however, expected to have a population increase by only 16 percent between now and 2040 (IMF, 2018). The impact of the demographic dividend is relevant. For example, job creation largely outpaced population growth in Angola, Botswana, Congo, Ghana, Mali, Senegal and Zimbabwe (IMF, 2018). However, in Burundi, Cote d'Ivoire, Kenya, Lesotho, Mozambique and Sierra Leone, population increase between 2000 and 2017 outgrew job creation (IMF, 2018).

Demographic structure is essentially an important housing demand factor because age distribution impacts on real estate market dynamics. It informs investment decisions in the type, scope, spread and location of real estate assets. The combination of limited supply and high demand of residential properties will continue to spur economic growth in most SSA countries. In particular, the medium- to long-term outlook appears promising given the high percentage of population (41%) in the under 15 years bracket. The situation presents an opportunity to analyse the dynamic effect of demographic structure on housing demand and estimate long-term housing demand in different sub-regional markets. Invariably, it will provide useful information for the supply-side and guide real estate market reforms in SSA.

The impacts of demographics on real estate markets are several because real estate investors respond in a number of ways. First, they can create new accommodation such as apartments, detached, semi-detached, townhouses, retirement villages, care and nursing homes in response to high demand. Second, investors can refurbish existing space or convert buildings currently with alternative uses. Third, the trend of young population will support a growing student population in the continent that guarantees an increasing preference for occupying privately operated student facilities as opposed to house sharing or living in family homes. Fourth, investors can take a view on these demand drivers to inform decisions on residential investments in major cities. As it can be seen, real estate investors and professionals can become innovative in response to market dynamics such as demographic pressures. In some cases, this may be escalated to culminate in the creation of new real estate products.

Demographics further offer diverging trajectories of labour force growth, market shifts in resource allocation, and the need for new approaches to developing and changing workplace environment. There is therefore the need to establish a solid foundation to attract talent and also create opportunities for competitive real estate investments and occupiers in SSA.

6. Foreign direct investments

Sub-Saharan Africa as at 2019 hosted several of the world's fastest growing economies. Among these economies were Ethiopia, Rwanda, Ghana, Cote d'Ivoire, Senegal, Benin, Kenya, Uganda and Burkina Faso (AfDB, 2011). Prior to covid 19, the region was on the verge of achieving an overall economic growth of 3.8 percent, slightly above the global target of 3.7 percent. The region is also characterised by a huge emerging middle-income class, usually defined as an individual with annual income exceeding UD$3,900 in purchasing power parity terms (AfDB, 2011). The middle-income population has tripled over the last three decades to 313 million people, or more than 34 percent of the continent's population.

The pattern of middle-income category varies across the continent with the highest percentage in the North African countries. Tunisia has the highest concentration of 90 percent, followed by Morocco and Egypt with 85 percent and 80 percent respectively (AfDB, 2011). Ghana (47%), Nigeria (23%), South Africa (43%), Namibia (47%), Gabon (75%), Cape Verde (46%), Botswana (48%), Kenya (45%) and Cote d'Ivoire (37%) are the other countries with high concentration of middle-income class (AfDB, 2011). This situation has yielded an increase in the size and purchasing power of the people. It is also underpinned by a strong economic growth and a move away from the historical peasant farming towards stable and salaried job culture. These stipulations together with the ease of doing business ranking of the World Bank continue to present strong signals in SSA for foreign direct investments (FDIs). Egypt, Morocco, South Africa, Nigeria and Kenya are in competition for most of the FDIs in the continent. Meanwhile, sources of FDIs into sub-Saharan Africa are traceable to five main countries – US, France, UK, China and South Africa.

It is expected that FDIs will focus specifically on job creation and long-term investments in infrastructural developments, retail, telecommunication, financial and technology to transform sub-Saharan Africa economies. Microsoft, the world's technology giant, for example, in 2020 committed to invest US$100.0 million in technology development centres in Nigeria and East Africa, possibly Kenya or Ethiopia, which is believed to possess great potentials in that part of the continent.

The size of each market in SSA is a matter of concern. Each government is obliged to roll out interventions and initiatives to support its real estate market. For instance, in Morocco, there is a deliberate government effort to make the country an automobile hub in the continent. Digitisation of the region's economies will also directly offer a conducive avenue for FDIs. This venture will nonetheless require massive investments in technology infrastructure such as TechHubs development. The African Continental Free Trade located in Ghana can be a game changer for the continent to stimulate growth in FDIs.

7. Remittances

A large number of nationals from the sub-Sahara African countries have migrated to wealthier countries or states across Europe, North America, Asia, the Middle East and within the continent; in South Africa, among others. Countries such as UK, US, France and China have contributed up to 25 percentage of all funds remitted to Africa. Gupta S. *et al.* (2007) argue that sub-Saharan Africa has been part of the increasing global trend with remittances increased by over 55 percent in US dollar terms since 2000, while they increased for developing countries as a group by 81 percent. Though recorded remittances in SSA are only a small fraction of the total remittances, it is estimated that informal remittances to the region are relatively high between 45 percent and 65 percent of formal flows, compared to between 5 percent and 20 percent in Latin America (Freund and Spatafora, 2005).

Remittances to 34 countries in SSA in 2005 were estimated at US$6.5 billion, and on average, 2.5 percent of GDP with the exception of Lesotho (28% of GDP) and more than 5 percent in Senegal, Cape Verde and Guinea-Bissau. In absolute terms, Nigeria, Senegal and Kenya were the largest recipients of remittances for the same period. The region has subsequently experienced a steady growth in remittances from US$29 billion in 2009 to US$48 billion in 2019, but predicted to decline in 2020 to US$37 billion (World Bank, 2020). The World Bank further forecasts that there will be 4 percent recovery to US$38 billion in remittances flows in 2021. The region emerged as one of the fastest growing regions globally for remittances between 2017 and 2018.

The World Bank (2020) clarifies that the role of remittances varies greatly from country to country. Whilst Nigeria received US$23.8 billion in 2019, as nearly half of the remittances to the region, South Sudan recorded the highest remittances; 34.4 percentage of GDP. In general, it is more expensive to send money to SSA than to any other region in the world, though competition among service providers is making it more affordable over time.

Balde (2011) explains that migrants and households spend a share of remittances on investment goods, including housing, education and small business, which are expected to strengthen the human and physical capital of recipient countries. For example, an increase of 1 percent in remittances in 13 Caribbean countries led to an increase in domestic private investment by 0.6 percent relative to GDP (see Mishra, 2005). Remittances have reduced credit constraints in the receiving households in the Philippines, El Salvador and Mexico, and also encouraged entrepreneurship in these countries (see Funkhouser, 1992 for El Salvador; Yang, 2004 for the Philippines, and Woodruff and Zenteno, 2001 for Mexico). Adams (2006) justifies that households in Ghana have treated remittances as any other source of income and there is no disproportionate tendency to spend it on consumption.

A survey conducted by the *Comité Français pour la Solidarité Internationale* (CFSI) and published by the *Agence Française de Développement* AFD in 2004 elucidates that migrants from Mali, Senegal, Morocco, Comoros and Vietnam living in France have classified their different motives to remit by importance and priority in this ranking: first, to assist family back home; second, real estate development; third, support community infrastructure development including health services and school; fourth, start a business; and fifth, open a saving account. An investment in real estate assets is surely one of the competing uses of remittances in SSA.

Singh *et al.* (2009) proves that remittances are larger for countries with a larger diaspora or when the diaspora is located in wealthier countries. Although the effect of remittances in growth regressions is negative, the study shows that countries with well-functioning domestic institutions seem nevertheless to be better at unlocking the potential for remittances to contribute to faster economic growth. Remittances therefore play an important role in stimulating growth in real estate markets and national development.

8. Property technology

The application of information technology and platform economics to real estate markets is popularly referred to as property technology (PropTech). It is an evolution of modern technological interventions in real estate markets that focuses on information, transactions and management. PropTech seeks to provide digital platforms to improve tenure security, registration and recording of rights, property valuation, plans preparation, and data analysis and retrieval. It can therefore be employed to facilitate real estate transactions, management and research. In effect, it uses digital innovative products to address the needs of real estate markets by offering solutions to increase efficiency and cost savings, and an opportunity for effective management.

PropTech also targets digitalisation of real estate markets. It is fast advancing with countries such as UK and US that are classified as global champions. Though PropTech is only one small part of the wider digital transformation, it has evolved as the future of real estate markets. It challenges the sector to embrace skills training and culture change, as much as technology shift and evolving demands of market participants.

PropTech is accordingly making an impact in residential and commercial submarkets, with digitisation rapidly becoming a part of the broader technological transformation that is driving real estate markets as well as the financial sector. The rapid adoption of technological innovations is an evidence that real estate markets are ready and perhaps in need for a change. It seems to be a total deviation from the traditional belief that the sector is resistant to change. As more real estate markets begin to develop digital innovations, they will positively influence attitudes of market players and key stakeholders.

Across sub-Saharan Africa, digital technologies are making an impact in diverse ways including kids learning to code outside Niger's capital; drones delivering medical supplies to remote communities in Ghana, Rwanda and Sierra Leone; and mobile money interoperability (non-banking money transfers). Very soon real estate markets in most SSA will no longer be described as a sleeping giant in terms of digital transformation. However, there are still obstacles to digitising some real estate markets in the region. For example, a number of SSA countries have less reliable internet connectivity and limited electricity supply. Though the gap is narrowing rapidly, SSA lags other regions in internet penetration. In addition, mobile download speeds in the region are, on average, more than 3 times slower than in the rest of the world. Affordability is also a lingering obstacle to adoption as the cost of accessing digital technologies remains high relative to incomes (IMF, 2020).

Property technology that provides digital platforms to improve land administration systems, develop real estate market and stimulate real estate investments is certainly at its infant stages in most SSA countries. Within each country's broader development agenda, there is the need for a digital strategic plan to massively invest in: sustainable digital policy and regulatory framework to foster a digital-friendly business environment; education to develop information technology skills; risk management to curb cybersecurity; and developmental infrastructure to include energy for reliable and affordable electricity and information technology. An investment now in a digital Africa will pave the way for more resilient SSA economies tomorrow to the benefit of their real estate markets. Governments, academia, practitioners and all other stakeholders therefore have a key role to play in that process.

9. Mortgage market

Understanding the real estate market's dynamics and trends in homeownership, real estate investment and real estate finance have become increasingly important in national development. In particular, a household's financial status and its future expectations are factors that influence the decision to rent or to own a residential property. Dozens of countries in Africa are faced with housing deficits, with demand far in excess of supply. Causes of the deficit include lack of capital, cost of capital, land tenure insecurity, overstated housing demand, housing delivery inefficiencies and infrastructure constraints. A few other country-specific challenges also exist.

Housing affordability remains a challenge across the region. The high cost of housing construction due to overdependence on imported materials such as iron rods, cement, floor and wall tiles, plumbing, electricals, and other fittings and finishes could be the primary cause. The gravity of the impact of housing affordability on national development thus depends on the country's economic environment and its government's specific interventions. Poverty persists in many countries in SSA with most people engaged in the informal sector and agriculture. In most countries in the region, the private sector drives the delivery of housing to middle- and upper-income brackets. Housing delivery for the low-income group accordingly becomes the burden of the state. Governments must endeavour to implement sustainable housing policies that provide adequate finance and build the capacity to deliver houses to meet the general market demands.

Mortgage markets across SSA, with the exception of South Africa, Mauritius and Botswana, are generally weak, and at the infant stages of development. Long-term mortgage vehicles, for example, do not exist in most countries. The absence of pension funds and insurance companies in the mortgage markets may have accounted for the dominance of few mortgage and commercial banks in the housing finance markets in most countries in SSA. For this reason, access to mortgage finance is hampered by weak credit systems. Regulatory frameworks for collateral and insolvency are commonly weak whilst credit history records hardly exist.

Whether real estate is treated as housing and part of social development or increasingly as an investment vehicle, the role of the state and financial actors is essential. Countries across SSA do not have a choice than to develop sustainable long-term mortgage markets. Governments should play that unique role not only in establishing real estate finance systems, but also providing a functional regulator. A policy driven action for pension funds and insurance companies to participate actively; establishment of mortgage banks and housing development agencies; and the provision of fiscal incentives for the private sector are strategic interventions that each country requires to develop its mortgage market. This will ultimately promote real estate investments in the sub-region.

10. The novel corona pandemic

The World Health Organization on March 11, 2020, declared the outbreak of corona virus (covid 19) as a global pandemic. Covid 19 pandemic has, and continues to impact many aspects of our daily life and the global economy, with some real estate markets having experienced lower levels of transactional activity and liquidity. Its adverse impacts have significantly affected not only sectors such as commerce, tourism, finance, health, education and manufacturing, but also real estate. Many SSA countries are characterised by a high degree of informality that places their economies at a high risk. For example, the evidence that 7 in every 10 of the total people employed in Ghana are in the informal sector (GSS, 2019) poses a severe welfare challenge on individuals and households. More seriously, a higher fraction of the total people employed is considered to be in vulnerable employment (66.2%) and underemployment (21.4%), pointing to a clear incidence of relatively low earnings among Ghanaian workers (NDPC, 2020). Major cities around the world, SSA not exempted, have borne the brunt of covid 19. The most affected areas in SSA are urban cities due to their uniqueness with high population, sizeable slums, large sizes and human connectivity (within and outside). The covid 19 pandemic has also exposed the basic flaws in urban planning and management, and public health systems globally.

The pandemic is not necessarily a typical real estate problem, but the demand and supply dynamics of many real estate markets may have been affected. Overall, the impact on real estate markets is relatively manageable because these are not as sensitive as the stock markets. Sales and rentals have slowed whilst each country's underlying demand and supply factors have rather determined the margins and trend. Needless to say, real estate is generally a more stable asset class. It is therefore expected that the markets in SSA may not be seriously affected by covid 19. It is thus anticipated that real estate markets in the region will survive in post covid 19 with stronger market performance indicators than other investment vehicles.

Lessons from historical antecedents, such as the great recession of 2007 to 2009, will inform current and future responses to the covid 19 pandemic. The economic downturn which exceptionally resulted in the bursting of the US housing bubble and global financial crisis had a slow recovery, up to five years to recover. Real estate markets suffered some consequences including mortgage contracts rigidity, sanctions associated with foreclosure, removal of incentives on debt repayments and high cost of long-term capital.

The effects and aftermath of the covid-19 pandemic will surely dictate the work of professionals in the built environment, particularly valuers in many ways. Government-imposed restrictions, company's internal arrangements or the occupant's unwillingness to grant access will definitely limit property inspections and valuations. Restricted access to property data such as comparables in emerging SSA economies may further be less freely available. This situation should compel professionals to conduct themselves in a more transparent and professional manner. The Royal Institution of Chartered Surveyors (RICS), for instance, has accordingly cautioned its regulated members to be aware and act within the requirements of the RICS Valuation – Global Standards (Red Book Global standards) with specific reference to: VPS 1.3.1 (i), which dwells on the nature and extent of valuer's work including investigations and limitations thereon; VPS 2.1 on inspections and investigations; VPS 3.2 (g) on extent of investigations; and VPS 4.8 on assumptions.

Appropriate and sustainable policy responses by governments and professional bodies should determine the future path for real estate markets in SSA. There should be an urgent need for governments' specific policies and regulations on real estate to address: mortgage financing with particular focus on limiting banks from starting foreclosure proceeding as much as possible; tenure security including eviction moratoria; technology driven platforms including digitisation; and sustainability of real estate development by providing incentives such as stimulus packages.

11. Conclusion

A fully functioning and well-regulated real estate market is undoubtedly relevant and a potential contributor to national development. Governments in the sub-region should therefore ensure that appropriate policy and regulatory frameworks are put in place in their respective countries to unleash the potentials in the African real estate markets. An integrated national development agenda that encompasses massive infrastructure development in telecommunication, information technology, energy, roads and railways; ease of doing business; security of lives and properties; certainty in policies (so that a change in government does not affect policies of previous governments); and strict compliance to the rule of law is indeed fundamental for every real estate market in Africa.

The state has a significant role in how patterns of real estate investment unfold, more so with its urban growth and land use management practices. Urban growth management has become an important tool for national, state, and local governments to respond to the growth and associated challenges facing African cities, generally through development legislation and planning policy, spatial development frameworks, and land use management regulatory regimes. O'Sullivan (2003) argues that in the context of market driven property economies, we find that urban and land use growth management practices that enable spatial patterns of property values to gradually increase from one segment to the other through filtering do facilitate greater equity and inclusion in the property market. However, if property markets operate at extreme ends, then relatively high values catering for the few elite's vis-a-vis the majority captured within lower segments of the marginal values, resulting in industrial scale exclusion of the poor urban majority. African cities struggle to devise approaches that are appropriate to the contextual realities of the continent. The patterns of inclusion and exclusion are not recent and can be traced back to ancient and colonial times as reflected in table 3.1 and discussed in this chapter. Watson (2006) argues that unless urban growth management practices and patterns of real estate investment can converge on the critical challenges facing African cities, the social, spatial, and economic structural problems will not be addressed and will rather deepen.

Ancient Settlements, Settlement Growth, and Governance

Neolithic settlements in Africa can be traced back 3500 BC to settlements in North, West, and Southern Africa such as the Ancient Nile Valley settlements, Yoruba cities, Ibadan, Ogbomosho, Iwo, and Ife. North Africa has been ruled and fought over by Phoenicians, Egyptians, Greeks, Romans, Islamic Caliphates, Ottoman Sultans, British, French, and Spanish colonists. Established kingdoms in West Africa (e.g., Mossi and Akan kingdoms) founded settlements such as Ouagadougou (Burkina Faso), Kano (Nigeria), Dagomba (Ghana), and Timbuktu (Mali). East Africa settlements (for example Zanzibar and Dar es Salaam) had strong historic links with Arab and Asian kingdoms settlements. Settlements in Southern Africa include Great Zimbabwe, the Cape Khoi and San of the Kalahari Desert, and surrounding regions; the Bantu and Nguni people and Zulu Kingdoms along the eastern and southern Africa coastal regions (Mumford, 1961; Kostof, 1991; Freund, 2007; UN-Habitat, 2013, 2014). Freund (2007) with his seminal work, *The African City: A History*, observes that there are or were uniquely African kinds of cities and towns, before the incorporation of Africa into world-systems with wide economic networks and defining urban cultural structures, as a cultural statement of Africanness.

The effect of all the diverse and unique forces on the settlements shape, size, and design is very difficult to untangle. Some of these earlier settlements around the African continent generally had a core, were walled, and were clearly separated from the surrounding farming and rural areas. They grew organically and although class, race, and income differentiation and inequalities existed, their physical design and structure allowed for the poorest residents to benefit from significant public investment in infrastructure by merchants, traders, ruler, regents, and royals (Mumford, 1961; Kostof, 1991). This is evident with the extensively documented role and function of public squares as multifunctional, trading markets and public roads, by all those who reside in these ancient cities. Public spaces were located and designed in a manner that did not purposely exclude poor residents to partake and engage in the social,

TABLE 3.1 African Cities: Urban Growth, Form and Structure; Urban Growth Management and Patterns of Real Estate Developments

Epoch		*Urban Growth, Form, and Structure*	*Urban Growth Management Practices*	*Patterns of Real Estate Developments*
Ancient	BC 1400/1500	• Walled settlements common. • Historic core, separation from rural areas. • Organic settlement growth.	• Perspective of settlement as holistic entity.	• Integrated authority, landholder, structure development. • Dominance of authority investment.
Initial colonial settlements *Scramble for Africa* *Apartheid in RSA*	1600s– 1800s–1900s 1948–1994	• New settlements established for extraction of natural and human resources (slavery). • New urban cores established. • Ancient settlements ignored and undermined.	• Attempts to contain urban settlements. • Limits to urban expansion due to infrastructure and transport constraints. • Colonial constructed viewed as legal and indigenous as illegal.	• Significant investment in public spaces, emulating colonial cities. • Private and company property investment within new urban cores.
Independence/ Self-governance	1930s–1970s	• Continuation of colonial cities. • Inner city public housing programmes, influence by Keynesian economics. • Importation of suburbanisation and "modern" new town concepts.	• Importation of land use control systems e.g., zoning schemes from the West UK. • Notions of "formal and legal" contrasted with "informal and illegal".	• Continued public and private sector investment in inner city areas. • Public housing on outskirts of the city.
Neo-liberal Influence & Structural Adjustments (Bretton Woods Institutions)	1980s/1990s onwards	• Dissection of the urban form and structure. • Urban seen as discrete elements that can be separated and delinked.	• Economic collapse and instruction from IMF and World Bank for governments to recoil from public sector investment. • Compact, integrated city policies emerge, with resistance to implementation. • Informal and illegal viewed as transitional – clampdown on "informal and illegal".	• Wave of private sector flight from cities. • Laissez-faire developments mainly through foreign investors. • Public sector investment minimised, both in terms of construction and maintenance.

(*Continued*)

TABLE 3.1 (Continued)

Epoch		*Urban Growth, Form, and Structure*	*Urban Growth Management Practices*	*Patterns of Real Estate Developments*
Contemporary Resurgence of Neo-liberalism interrupted with economic recessions and pandemics.	Post 2007 Recession	• Unstructured, unplanned city. • Footloose gated estates and new towns for the minority wealthy, preference for prime farmland or natural settings. • Informal settlements, squatter camps, inner city ghettos for the poor urban majority. • Critical urban infrastructure (e.g., transportation routes serving formal capital and the property elites).	• Sustainable, integrated cities in rhetoric and active resistance from the property elite. • Some degree of land use control, however ignored by both wealthy and poor residents. • Continuation of approach to informal attempts at formalisation of "illegal" settlements and trading.	• Separated high, medium, and low value property investment. • Private investment based on non-public sharing of communal facilities. • Public investment prioritising the needs of the wealthy (e.g., private road infrastructure).

Author constructed. Sources: Dewar et al. (1986), Gore (1984 and 2000), Hart (2001), Todes and Harrison (2001), Slater (2004), Peet and Hartwick (2009), Barca et al. (2012), Scott (2012), Martin (2015).

economic, and political life of the city. The governance of these ancient settlements would have occurred through the ruling powers with no notion of "legal and illegal" with regard to property. The "legality and illegality" of property came in with colonisation.

Colonised African Cities, Settlement Governance, and Real Estate Investment

The onset of colonial occupation of Africa by the Europeans during the 17th century ushered in the start of large-scale extraction and exploitation of the continents' human and natural resources, such as agricultural produce, fresh water, minerals, cheap labour, and human slaves, spearheaded by mega corporations such as the Dutch East India Company (VOC), officially the United East India Company, during the 17th century. The initial colonial settlements and the "Scramble for Africa" occurred during the 18th and 19th centuries, with France and Great Britain colonising West Africa; Great Britain, Portugal, and Germany colonising South and Eastern Africa; and Belgium and France colonising Central Africa. Ethiopia was the only country not colonised.

It was common practice with this wave of colonial invasion to establish new trading and refreshment stations away from existing traditional villages and towns, and construct the new settlements primarily as defensive enclaves with the grid as basic design, in accordance with urban places in Europe (Freund, 2007; UN-Habitat, 2013). Freund (2007: 60) observes that "it was decommissioned Dutch East India Company employees and other immigrants who came to provide the key services that the economy required. As land loans in the vicinity of the castle gave way to outright land purchase". The term "real estate" was initially recorded during the 1600s and was derived from the Latin meaning of "actual" or "genuine", while estate refers to "land", real estate refers therefore property consisting of houses or land. The settling of large numbers of colonisers in African cities resulted in the management of growth of urban settlements in a manner that ensured that the areas they developed were the historic core and were reserved for the colonial elites. In some cases, the poorer indigenous population were accommodated within these settlements, typically within less desirable and more industrial parts where labour was needed. In these areas, they were not allowed to "own" the land they occupied, but they were renters to leave when their labour was not needed. "From the late nineteenth century, segregation was the colonial watchword throughout the continent. The colonial city emerged and differentiated itself, often ruthlessly from pre-existing patterns" (Freund, 2007: 78). These activities represent the first "land grab" for speculative and profit purpose on the African continent by the colonisers. The management of urban growth included attempts to contain the settlements, with limits to urban expansion due to infrastructure and transport constraints. The state pursued inner city public spaces and housing programmes for their benefit.

Independence, Urban Growth Management Response, and Real Estate Development

The colonial established centres became the primary centre of governance for states with independence (Young, 2004; Freund, 2007). Parts of these settlements developed by the colonisers, including both quarters for the wealthy and the poor, were regarded as "formal and legal". In contrast, efforts by local, mainly indigenous populations to establish settlements on the periphery of urban opportunities, which to a large extent took the form of shanty towns, were regarded as "informal and therefore illegal". These patterns of differentiation between "insiders and outsiders" to the property market were retained and further entrenched within these settlements, which for decades were initially relatively small, grew organically and expanded immensely over the last century, some into towns, cities, and metropolises. The construction of centralised public spaces and facilities and public housing were to a large extent continued by the newly independent states and cities. This period was characterised by the importation of land use control systems e.g., zoning schemes from the West (the UK and the USA in particular). Notions of "formal and legal" contrasted with the identification of what was regarded as "informal and illegal".

The influence of colonial powers on African cities is most prominent through the importation of the "modern" new town and suburbanisation planning models from the US, UK, and Europe. Their importation to manage urban growth in African cities aggressively during the 1960s to 1990s provided a useful "technical" approach to retain and entrench spatial separation between "elite and lower classes" (Badibanga, 1985). These included suburban neighbourhood designs, such as Garden Cities (UK, Ebenezer Howard, 1902), the Neighbourhood unit (USA, Clarence Perry, 1920), the New Town Movement (UK, Melvin & Walker, 1967), Woonerf concepts (Netherlands, 1969), from the USA, UK, and Europe and ushered in industrial scale townships developments for the mainly poor, indigenous urban residents.

In South Africa it was accompanied by the brutal forced removals of non-white residents during the 1960s to 1980s from inner city areas with the full-scale implementation of apartheid. These applications are particularly evident in former British colonies such as South Africa, Zimbabwe, Kenya, and Zambia through land use planning mechanisms and zoning scheme regulations (Freund, 2007; UN-Habitat, 2013). Although African cities have diverse physical designs and functions, they all tend to display similar urban form features and these include in broad terms, formal forms of development and informal forms of development. A prominent feature of cities in Francophone countries is the existence of "enclavements", as a consequence of the "fragmentation of the colonial city, . . . the exclusion of the masses from central facilities . . . and the absence of decent public transport systems" (Freund, 2007: 86). These patterns provided the spatial, economic, and political context for the planning, design, construction, and pattern of real estate investments. Formal forms of developments include typically the historical downtown, the seat of the colonial, commercial, and financial powers; suburbs for the middle class (generally with large residential plots and public amenities such as schools, clinics, and other amenities); and townships for the majority and poorest residents, generally located outside the urban core and underserved with basic public utility services and facilities. These areas developed into very different and apposing real estate markets and are underscored by the myriad of problems and challenges facing African cities. The primary drivers of formal real estate development during the colonial and immediate post-colonial periods were governments (Konadu-Agyemang in Awuah et al., 2015). The state was the primary driver of residential development prior to the 1980s, motivated primarily by political and social welfare considerations. However, the informal sector formed the largest proportion of residential development on the continent as was the case during the colonial and immediate post-colonial periods (Arku, 2006; Freund, 2007; Centre for Affordable Housing Finance in Africa, 2010 in Awuah et al., 2015).

Neo-liberal Influence, Urban Growth Management, and Property Investment

Structural adjustment programmes were conditional loans introduced by Bretton Woods Institutions such as the World Bank and International Monetary Fund (IMF) during the 1980s and 1990s for highly indebted, low-income African countries. These included many sub-Saharan countries in Africa such as Ghana, Zambia, Cote D'Ivoire, Senegal, Uganda, and Zimbabwe. These programmes required the introduction of austerity measures by the state in particular on social spending, in particular on health care, education, and public housing, coupled with the privatisation of public assets (Awuah et al., 2015). Awuah et al. (2015: 124), reflecting on experience in Ghana, note that this "introduced pro-market ideas and succeeded in linking the continent to the international capitalist production system and the state withdrawing from direct involvement in housing finance and production". The structural components of the city, in effect the infrastructure, public spaces, and public housing, were discrete, separate elements that can be delinked and commodified. The uncoupling of cities in Africa through the insertion of these neo-liberal forces was accelerated with the importation of suburbanisation and land use separation and control measures once more, from the west.

The property development sector responded with waves of private sector flight from cities driven by laissez-faire developments mainly through foreign investors. Public sector investment was minimised, and in some cases ceased, both in terms of construction and maintenance of infrastructure and public facilities. Consequently, "formal private sector investment in real estate development is now a prominent feature of Africa's real estate market" (Awuah et al., 2015: 124). Formal private sector real estate developments generally take the form of gated communities, serving the needs of the middle and upper class on the outer fringes of the main urban centres in cities such as Johannesburg, Lagos, and Accra (Centre for Affordable Housing Finance in Africa, 2010 in Awuah et al., 2015). The developers are mainly foreign investors, focussing on purchasing large tracts of land on the outskirts of cities. The benefits of these developments include an increase in housing stock and infrastructure and sense of security in the gated communities. However, these developments only serve the middle and upper income groups and create urban segregation between the rich and poor (Awuah et al., 2015). Regarding urban growth managed responses, the recoiling of state investment in infrastructure reduced the ability of the public sector to influence the pattern of private sector investment. Although progressive policies emerged on compact, integrated, and sustainable cities, this was met with resistance to implementation, mainly by vested interest in the private sector. Both formal and informal developments do not comply with planning and urban regulations and are resulting in urban sprawl (Awuah et al., 2015).

City governments in Africa have retained the land use system imported from their former colonial masters, undoubtedly to secure property rights that were introduced with colonialism and further entrenched with the imposition of neo-liberalism and structural adjustment as well as national constitutions, such as South Africa's. Many formal colonies have planning ordinances and systems based on the British Town and Country Planning Act that are complex and bureaucratic requiring lengthy approval processes such as zoning, approved subdivision, and building plan permits from planning authorities before development is authorised (Freund, 2007; Awuah, 2015; Madell, 2016). This approach to land use management reinforces a market orientated approach to property development and actively differentiates between "formal" and "informal" property, i.e., real estate that is accounted for within legal and tax systems and that which is not. In practice, this approach excludes the welfare of the majority of urban residents in cities in Africa as they are unable to comply with "formal" standards, especially as far as the building regulations are concerned.

The informal housing stock has been constructed mainly by private individuals on an incremental basis on land acquired from customary landowners through leasehold agreements (Konadu-Aqyeman, 2001; Grant, 2007 in Awuah et al., 2015). Most of these informal sector owner-developers do not comply with planning requirements or develop in unplanned areas. It is common for these developments to take a long time to be completed and are therefore often incomplete, deserted, and at different stages of construction (Awuah et al., 2015). In Algeria, during the first years of independence (early 1960s), poor rural migrants who established squatter settlements on the urban periphery, and therefore in defiance of modernist planning, were forcefully removed (Freund, 2007). In South Africa and Zimbabwe, clampdown on "informal and illegal" land occupation and construction of trading and housing structures is a widespread, endemic, and contemporary phenomenon.

Awuah et al. (2015) argue that the continents land use planning regimes are a disincentive and constraint for real estate investment. The authors cite examples in Ghana, where formal property developments are in effect "illegal" as approval has not been granted by relevant planning authorities. The authors note that the sprawling nature of formal developments is in breach of planning and urban development regulations, attributed to the country's weak planning system that levies a significant cost on property developers, and therefore has an adverse impact on the development return. The authors conclude that "there is therefore a need to review the existing planning arrangements, to create the appropriate mix between planning ideals and formal private sector development requirements to provide incentives for real estate developers" (Awuah et al., 2015: 134).

Contemporary Urbanisation, Urban Growth Management, and Property Development

Contemporary patterns of urbanisation and modernisation in African cities are taking a particular form, given its colonial history and tendency to follow trends that have emerged in developed countries, albeit trends in real estate developments tend to lag these patterns by a decade or two (Madell, 2016). A wide range of global trends is already evident in the African real estate industry, while others will only start to appear in the future (ibid). The pattern of skewed access to wealth in favour of the few, is evident in the highly disparate access to and control of fixed assets such as property. The control and main features of the real estate markets in African cities represent critical forces at play and represent, in broad terms, "formal and legal" interests of established elites that can be tracked back to pre- and post-colonial interests; and "informal and illegal" interests of the urban poor in response to the barriers to entry and opportunities and that arise from the interplay between economic, social, and political forces at play.

Most countries in Africa have retained the core of their colonial inherited growth management policies, land use ordinances, and related legislation, even though these have not been able to respond to the contemporary challenges facing African cities (Smit, 2018). Urban growth management has become an important tool for national, state, and local governments to respond to the growth and associated challenges facing African cities, generally through a combination of forward planning and land use management. The former has taken the form of masterplans, structure plans, spatial developments frameworks, neighbourhood designs, and informed primarily by policy and to a lesser extent by legislation. This, while the later centres on zoning schemes and associated land use regulations, informed mainly by legislation, municipal bylaws, and national building regulations. These are generally managed at a municipal level and include mechanisms such as rezoning, subdivision, departure, consent use, site development plan, removal of restrictive title deed conditions and administrative penalty procedures, and applications. The origin of efforts at managing the growth of settlements can be traced back to earlier periods of colonisation and to a large extent favoured profit orientated "formal and orderly" developments (Awuah et al., 2015).

The form and impact of urban growth management have been central to the problems facing African cities, given the colonial inherited design and function settlements and relatively low levels of public and private sector investments. Towns and cities in Africa have pursued imported and locally devised urban growth management practices that they believe will respond to the myriad of development challenges.

> Many African cities are unable to manage rapid urbanisation using static blueprint master plans, insufficient and inadequately trained planning staff and other resource limitations; consequently, towns and cities cannot cope and this can be attributed to the limitations of master planning with its assumption that cities will be dominated by

> formal developments, as in the case of cities in western countries; in cases where city management and duties have been decentralised, this has not been accompanied by power and resources to drive these processes.
>
> *(UN-Habitat, 2013: 17)*

A key feature of managing urbanisation in Africa is the dualistic focus on rural areas. This often contradicts efforts by the state to mobilise resources for urban development and to implement strategies aimed at creating compact and denser cities. Consequently, dispersed forms of settlements that are more compatible with informal settlements are encouraged, and this undermines efforts at consolidating urban growth and reducing the cost of providing much required bulk infrastructure (Turok, 2013; UN-Habitat, 2013).

High to Medium Value Formal Real Estate Investment and Urban Management

The formal and established face of Africa's real estate markets historically displays similar features to those in developed economies in the west, although lagging in terms of trends and generally at significantly lower levels of investment in comparison with global patterns. Real estate investment in African cities has been varied, with the public sector in most cases pursuing developments independent from the private sector. Private developments have taken a laissez-faire form and inform by market forces, often with tacit political support within the guise of goals such as to create a "world class city" and "make cities safe for capital" (McDonald, 2008). These developments included predominantly decentralised nodal and suburban developments, shopping centres, local retail nodes, exclusive high-income neighbourhoods, gated communities, golf and wine estate developments, high value industrial estates, warehousing, call centres, and logistic hubs, private sector driven. These forms of urbanisation in African towns and cities demonstrate widespread suburban sprawl evident through low density and "leap-frog" developments, in particular on the edge of settlements. These are generally located away from low income- neighbourhoods, and prefer areas that provide high mobility transportation infrastructure, utility service and logistic support, and neighbourhoods with more skilled labour force. These forms of urban development are apparent with major real estate projects in cities on the African continent such as Kigali, Lagos, Accra, Dar es Salaam, Nairobi, Kampala and Johannesburg, and Cape Town and reflect continuation of a tendency to import development and ideas from the west and more recently the far east. PWC (2015) predictions are that fast-growing cities will present a wider range of risk and return opportunities; the global investible real estate universe will expand substantially, leading to a huge expansion in opportunity and intensification of competition for prime assets will intensify; technology innovation and sustainability will be key drivers for value; and collaborating with governments will become more important.

A more concerning pattern of real estate investment is the visualisation of high-technology satellite towns, and the illusionary redevelopment of capital cities has been a more recent phenomenon. Watson (2014) referred to these as "fantasy cities", and examples include Waterfall City (Johannesburg), Tatu city (Nairobi), Eco city (Lagos), and Appolonia (Accra). Visions of these new cities are mainly driven by the private sector, with political support and promise of major foreign investments and catering mainly for higher income earners. These projects have been criticised, especially regarding their elitist undercurrents – in some instances,

regularisation of title may indeed inhibit the informal property market or move gains to the rich instead of those already inhabiting the land. In some cases, "sales are . . . more frequent when people lack legal title" (2002: 1). Roy (2001) questioned De Soto's assumption that once informal assets were legalised, "capitalist prosperity will flow into every corner of the world" (2001: 148) and concluded that De Soto's hypothesis, that the end of poverty was just a matter of legally recognising the trillions in assets the poor held, was "ludicrous" (2001: 152). Roy argued that a fundamental issue at stake in informality is that of wealth distribution and unequal property ownership. Formalisation of tenure does not always result in the desired outcomes of empowering the poor and bringing equity to property markets. In reality most poor residents would not risk their dwelling to secure a business loan due to its perceived high social and utility value (Finmark Trust, 2004 in Madell, 2012).

The urban management responses to informality from city authorities have varied from some degree of tolerance, for example in peri-urban areas in Zambia and Nigeria, to forced removals in South Africa and Zimbabwe. This approach to informality has been coupled with a recoiling of state involvement in providing urban infrastructure beyond basic utility services, and even to the extent where this is limited to temporary "settlement", as is the case in South Africa. In contrast, the state and local municipalities can adequately secure funding for the upkeep of bulk infrastructure (e.g., national and provincial routes), well before upgrading is required.

Urban Growth Management Practices and Real Estate Activities: Quō vādis?

Considering the patterns of growth, urban form, structure, and multitude of development challenges facing African cities, urban growth management responses (or lack thereof) have a considerable impact on the resultant real estate market activities. Ancient settlements evolved organically and were more inclusionary in terms of their characteristics and function, while colonial and contemporary African cities have evolved to primarily serve the interest of the few (Freund, 2007; Abubakar and Doan, 2017). The state in collaboration with the private sector have in effect developed a two-tier system, the formal and informal, both existing in a manner that is exclusionary and often contradictory.

The form of urban management deployed in many African cities during the colonial and subsequent independence period has generally served "formal and legal" developments. Urban growth management practices such as land use management systems have, to some extent, boosted the real estate development prospects of cities, evident with the entrenchment of property rights through municipal by-laws, development management, and zoning schemes (Charlton, 2012). Market driven real estate investments have generally succeeded in creating high value investments through upmarket housing estates, commercial, retail, and office developments (Todes and Robinson, 2020). These generally reflect the interest of the minority "formal" property owners in society, with a disproportionate amount of public resources allocated towards upkeeping and entrenching the system of land use registration and management (ibid).

It is common practice with runaway, laissez-faire, private sector driven real estate investment to use loopholes in the planning and urban management systems to depart from public sector determined values and goals articulated with land use management and spatial planning plans (Awuah et al., 2015; Madell, 2016). Real estate investments in townships, suburbs, and gated communities located on the outskirts of the metro and of surrounding towns, by both the public and private sector, have generally polarised and further entrenched colonial and post-colonial patterns of urban inequities across cities in Africa (Freund, 2007; Todes and Robinson, 2020). Ironically, market driven real estate markets have also thrived where they have chosen to ignore formal planning systems, as in the case of Ghana, where this avoidance is attributed to undue property development costs escalation (Awuah et al., 2015).

The state's role has regressed from securing the interests of the broader public and ensuring the poor is not excluded, to serving the primary interest of capital and the formal (Awuah et al., 2015; Efroymson, 2015). In many cities this has been accompanied with the destruction of natural systems and biodiversity, and the loss of productive and prime agricultural land, thereby undermining food security and accelerating the pace and impact of climate change, with increased hazardous risks for human settlements (Turok, 2013; UN-Habitat, 2013; UN-Habitat, 2014). The increasing costs associated with providing bulk infrastructure and public transport within the context of sprawling patterns of dislocated developments undermine urban growth and management efforts aimed at achieving affordable, efficient, and inclusive patterns of developments. The creation of mono-functional, single income group, dormitory suburbs and low-density developments are unsustainable socially, ecologically, and economically. More so, these developments exclude the majority of urban residents from meaningfully participating in the property market, creating wealth, and improving their well-being. Urban growth management impacts on real estate development and is central to addressing the challenges facing African cities. What is therefore the way forward?

Freund (2007) reflects on alternative views on development in African cities. One approach is to view shack-dwellers and the unemployed as the "authentic builders" of African cities, in essence "development from below" and a "self-help" city.

> From this perspective, the city merely illustrated the general proposition rapidly gaining ground amongst aid donor and Western observers of Africa that the Africa state had turned swollen and corrupt and was more of a nuisance and a bully than a genuine contributor to development of any sort. The state provided ordinances and decrees that showed very little variation from colonial patterns except that the capacity to carry out such plans was no longer there. These plans were in some respects nefarious in the way the privileged some social groups and in other ways irrelevant to the real social processes at work in the city.
>
> *(2007: 156)*

Another approach advocated by the World Bank during the 1970s to 1990s included the preparation of site-and-service schemes on greenfield sites on the edge of cities as an inexpensive way to allow for urban growth and fitting the "self-help" ideology. However, this turned out to be too expensive as the required infrastructure did not materialise or was poorly maintained. Other approaches include localised, autonomous local agencies and governments as pursued in Francophone countries (Freund, 2007).

Dewar and Uytenbogaardt (1991), with their seminal work on *South African Cities, A Manifesto for Change*, argue for managerial and planning judgements based on humanist concerns. These authors argue that this requires focussing on growth patterns and structural characteristics of the city as whole, a stop to planning comprehensively and rather embracing complexity and establishing frameworks to guide decision-making in space and allowing maximum freedom of choice. Proposals to management practices advocated include compacting the city, integrating urban activities and uses, seeking continuity of urban development, allowing more intensive activities along movement routes, and viewing collective spaces and places as the basic building blocks of urban systems.

Some African cities, such as Cape Town, Johannesburg, Durban, Lusaka, and Harare, have formulated forward planning spatial development frameworks or master plans informed by some of these urban growth management ideas (Todes et al., 2010; City of Cape Town, 2012; City of Johannesburg, 2016). These include measures such as the demarcation of urban edges to promote urban densification and protect agricultural land; urban nodes to focus the concentrations of economic, commercial, industrial, and/or residential activities at points of high accessibility; activity corridors, to concentrate mixed use developments along public transportation routes; transit-oriented development to enable sustainable public transport; urban renewal and public open space systems incorporating natural areas, active and passive recreational areas such as sports fields, parks, detention ponds (Harrison and Todes, 2015). However, with a few exceptions, implementation of these strategies has been ineffective, more so in the case of South African cities with their entrenched apartheid city urban form and structure (Harrison and Todes, 2015; Madell, 2016).

Towns and cities in Africa have pursued imported and locally devised urban growth management practices that they believe will respond to the myriad of development challenges and result in more sustainable and inclusive cities. Many African countries have signed the declaration in support of UN-Habitat's Sustainable Development Goals (SDGs). New Urban Agenda advocates for well-managed and sustainable cities can be powerful tools to promote equity, social welfare, and shared prosperity (UN-Habitat, 2013, 2015, 2016, 2018). For example, SDG 11 states, "make cities and human settlements inclusive, safe, resilient and sustainable". UN-Habitat (2018) advocates a human rights-based approach and "good governance" to planning and development.

Campbell (2018) poses the view that the solution to building a better society lies in harnessing many small, bottom-up ideas and actions to shape neighbourhoods. He advocates "making massive, small change" through "radical incrementalism":

> in which a as series of small changes are enacted one after the other, resulting in radical cumulative change in the formulation of neighbourhoods. While you should have a shared vision of where you want to go, the incremental steps to get there necessarily shape the course. The feedback you get along the way will initiate, accelerate, and modify your next action. It is iterative and adaptive learning at its best – governments and people gain knowledge along the way, better informing their future actions.
>
> *(Campbell, 2018: 13)*

Keeton and Nijhuis (2019: 219) borrow from these ideas and advocate, when considering the development of new towns and gated estates in Africa, the notion of "adaptive planning . . . understood as planning that embraces social and ecological specificities and acknowledges the diversity of different contexts . . . to accommodate grass-roots organizations and embrace diversity among citizens".

Conclusion

African cities characteristically display the notion of a "tale of two cities", the duality of a part of the city designed and managed within the image of "world class" cities, gated communities, golf estates, and shopping malls emulating designs in the developed world, but serving the minority of residents, in effect the property elites. In contrast, parts of the city where the majority poor residents struggle to survive, living in squalor and with minimum services, underserved and dislocated from economic activity and jobs, and from necessary public health, education, and administrative services. Ancient and colonial cities were constrained by technological limitations, to dislocate themselves from those they regard as poor and economically inferior. The arrival of the colonial powers, racisms, and other forms of dominance on African cities, boosted by neo-liberal economic policies and discriminatory land use management practices, have set the stage for the current poverty traps, inequalities, and marginalisation.

The efforts of powerful elites have directly and indirectly influenced the growth and structure of African cities and by default, the patterns of real estate investment, within the context of compounding development challenges such as massive inequality, poverty, and economic marginalisation of the majority of the urban population. The brief historic overview of African cities shows that roots of the current realities can be traced back to the political, social, and economic forces at play; the growth, urban form, and structure of settlements on the continent; the associated pattern of property investment and development; and the urban management response from the public sector. The complex and integrated relationships between these factors were explored over the ancient, colonial, and contemporary epochs. This analysis shows that the interests of those in power, in this case the property elites, have a significant impact on how the city is structured and developed and how the state generally serves this interest, or at least does not confront the forces at play.

However, the growth of cities has also in modern times been influenced by development philosophies such a Keynesianism and Humanism that recognise that communities are poor not by accident but by design and therefore advocate the welfare of broader society. The effective and efficient management of cities is central to addressing these challenges facing African cities. Corruption within the public and private sector impacting on the growth and development of African cities is endemic and must be dramatically curtailed. City management should be decentralised to regional and local levels, accompanied by a massive increase in capacity and overall improvement of urban political leadership and administration. The management of the urban form and structure of cities is vital when dealing with the daunting development challenges facing African cities.

Forward planning remains a critical tool and the importance of clear spatial frameworks and strategies to manage cities at a macro scale and to guide the development of bulk utility infrastructure, mass-based public transportation systems, major public facilities, and spaces that serve the city as whole is recognised by African urban scholars (Dewar and Uytenbogaardt, 1991; Turok, 2013; Harrison and Todes, 2015). Perhaps the time has come to radically change the approach to land use management in African cities and abandon archaic zoning schemes. This required change should be driven by local agencies and role players and combined with more context specific and "adaptive" planning and property development interventions. Localised versions of "performance based" and "overlay activity use" tools may offer some insights on how this could be achieved. This will require real estate actors to accept development parameters beyond the conventional formal domains, including properties and structures that are currently viewed as "informal and illegal", but that do meet minimum norms of security, safety, and health. Campbell's (2018) call for "making massive small changes" within the context of a shared vision holds significant relevance for real estate investments and property development in African cities.

Bibliography

Abubakar, I., and Doan, P. 2017. Building new capital cities in Africa: Lessons for new satellite towns in developing countries. *African Studies*, vol. 76 (4), pp. 546–565.

Adams, D. 1994. *Urban planning and the development process*. Abington: Routledge.

Adams, D., and Tiesdell, S. 2013. *Shaping places: Urban planning, design and development*. London and New York: Routledge.

Amis, P. 1984. Squatters or tenants: the commercialization of unauthorized housing in Nairobi. *World development*, vol. 12 (1), pp. 87–96.

Arku, G. 2006. The housing and economic development debate revisited: economic significance of housing in developing countries. *Journal of Housing and the Built Environment*, vol. 21 (4), pp. 377–395.

Avis, W. R. 2016. *Urban governance (topic guide)*. Birmingham, UK: GSDRC, University of Birmingham.

Awuah, K. G. B., Hammond, F., Lamond, J., and Booth, C. 2015. Ghana. Impact of land use planning on real estate development return in a developing world context. In Squires, G., and Heurkens, E. (Eds.), *International approaches to real estate development*. London: Routledge.

Badibanga, A. 1985. Imitative urbanisation and the outward growth of African cities. *Tiers-Monde*, vol. 26 (104), pp. 849–859.

Barca, F., McCann, P., and Rodriguez-Pose, A. 2012. The case for regional development intervention: Place-based versus place-neutral approaches. *Journal of Regional Science*, vol. 52 (1), pp. 134–152.

Campbell, K. 2018. *Making massive small change. Building the urban society we want*. London: Smart Urbanism.

Charlton, S. 2012. *Inequality and economic marginalisation: The state of land use management in South Africa*. Urban Landmark.

City of Cape Town. 2012. *Metropolitan spatial development framework*.

City of Johannesburg. 2016. *Spatial development framework 2040*.

De Soto, H. 2000. *The mystery of capital. Why capitalism triumphs in the West and fails everywhere else*. London: Bantam Press.

Dewar, D., Todes, A., and Watson, V. 1986. *Regional development and settlement policy: Premises and prospects*. London: Allen and Unwin*. Republished 2018.

Dewar, D., and Uytenbogaardt, R. 1991. *South African cities: Manifesto for change*. Cape Town: University of Cape Town.

Durand-Lasserve, A., and Selod, H. 2009. The formalization of urban land tenure in developing countries. *Urban Research Symposium*, May 14–16, Washington, DC, World Bank.

Efroymson, D. 2015. *Beyond apologies: Defining and achieving an economics of wellbeing*. Dhaka, Bangladesh: The Institute of Wellbeing.

Finmark Trust. 2004. The township residential property market. Johannesburg. Unpublished.

Freund, B. 2007. *The African city: A history. New approaches to African history*. Cambridge: Cambridge University Press.

Gilbert, A. 2002. On the mystery of capital and the myths of Hernando De Soto: What difference does legal title make? *International Development Planning Review*, vol. 24 (1), pp. 1–19.

Gore, C. 1984. *Regions in question: Space, development theory and regional policy*. London: Methuen*.

Gore, C. 2000. The rise and fall of the Washington consensus as a paradigm for developing countries. *World Development*, vol. 28 (5), pp. 789–804.

Gulyani, S., and Talukdar, D. 2008. Slum real estate: The low-quality high-prize in Nairobi's slum rental market and its implications for theory and practice. *World Development*, vol. 28 (5), pp. 789–804.

Harrison, P., and Todes, A. 2015. Spatial transformations in a "loosening state": South Africa in a comparative perspective. *Geoforum*, vol. 61, pp. 148–162.

Hart, G. 2001. Development critiques in the 1990s: Culs de sac and promising paths. *Progress in Human Geography*, vol. 25(4), pp. 649–658.

Healy, P. 2010. *Making better places: The planning project in the twenty-first century*. Basingstoke, Hampshire: Palgrave Macmillan.

Huchzermeyer, M., and Karam, A. 2006. The continuing challenge of informal settlements: An introduction. In Huchzermeyer, M., and Karam, A. (Eds.), *Informal settlements: A perpetual challenge*? Cape Town: UCT Press, pp. 1–16.

Jenks, M., & Burgess, R. (Eds.). 2000. *Compact cities. Sustainable urban forms for developing countries*. London: Spon Press.

Keeton, R., and Nijhuis, S. 2019. Spatial challenges in contemporary African New Towns and potentials for alternative planning strategies. *International Planning Studies*, vol. 24 (3–4), pp. 218–234.

Kostof, S. 1991. *The city shaped: Urban patterns and meanings through history*. New York: Little, Brown and Company.

Lynch, K. 1984. *A theory of good city form*. Cambridge, MA: MIT Press.

Mabin, A., Butcher, S., and Bloch, R. 2013. Peripheries, sub-urbanisms and change in sub-Saharan African cities. *Social Dynamics*, vol. 39 (2), pp. 167–190.

Madell, C. 2012. Local economic development, business stimulation and poverty reduction in poor urban areas: The disjuncture between state policies and household survival strategies in the case of Heideveld. PhD Thesis, Cape Town.

Madell, C. 2016. Historic overview of urban development in the Western Cape and in South Africa and refection on progress on spatial integration and sustainability 2016. *Law Reform*, DEADP, WCG.

Martin, R. 2015. Rebalancing the spatial economy: The challenge for regional theory. *Theory, Politics, Governance*, vol. 3 (3), pp. 235–272.

McDonald, D. 2008. *World city Syndrome: Neoliberalism and inequality in Cape Town*. New York: Routledge.

Mumford, L. 1961. *The city in history. Its origins, its transformations, and its prospects*. San Diego: Harcourt Inc.

Myers, G. 2011. *African cities: Alternative visions of urban theory and practice*. London, New York: Zed Books.

Nationsonline. 2018. *Capital cities of Africa*. www.nationsonline.org/oneworld/capitals_africa.htm. Accessed April 5, 2021.

O'Sullivan, A. 2003. *Why housing is different. Urban economics*. New York: McGraw-Hill.

Peet, R., and Hartwick, E. 2009. *Theories of development. Contentions, arguments, alternatives*. New York: The Guilford Press.

Pieterse, E. 2008. *City futures. Confronting the crisis of urban development*. London: Bloomsbury Publishing.

Pieterse, E. 2010, August. Cityness and African urban development. In *Urban forum* (Vol. 21, No. 3, pp. 205–219) Springer Netherlands.

PWC. 2015. Real estate building the future of Africa. March.

Ratcliffe, J., Stubbs, M., and Keeping, M. 2009. *Urban planning and real estate development*. 3rd ed. London: Routledge.

Roy, A. 2001. Urban informality. Towards an epistemology of planning. *Journal of American Planning Association*, vol. 71 (2), pp. 147–158.

RSA. 2013. Spatial planning and land use management act (2013).

Saghir, J., and Santoro, J. 2018. Urbanization in Sub-Saharan Africa. *Meeting Challenges by Bridging Stakeholders*, Centre for Strategic and International Studies.

Scheba, A., and Turok, I. 2020. Informal rental housing in the South: Dynamic but neglected. *Environment & Urbanization*, vol. 32 (1), pp. 109–132.

Scott, A. J. 2012. *World in emergence: Cities and regions in the 21st century*. Cheltenham, UK: Edward Elgar.

Sihlongonyane, M., and Karam, A. 2004. The impact of the national housing capital subsidy scheme on the apartheid city: The case of Johannesburg. *The Proceedings of the National Planning History Conference*, Bloemfontein, South Africa.

Slater, D. 2004. *Geopolitics and the post-colonial: Rethinking North-South relations*. Blackwell.

Smit, W. 2018. Urban governance in Africa: An overview. *International Development Policy*, vol. 10. https://journals.openedition.org/poldev/2637.

Squires, G., Heurkens, E., and Erwin, W. T. M. 2015. *International approaches to real estate development*. London: Routledge.

Stubbs, M. 1962; Keeping, M. 2009. *Urban planning and real estate development*. 3rd ed. London: Routledge.

Tiesdell, S., Adams, D. 1954–2011. *Urban design in the real estate development process*. Chichester, UK: Wiley-Blackwell.

Todes, A., Karam, A., Klug, N., and Malaza, N. 2010. Beyond master planning? New approaches to spatial planning in Ekurhuleni, South Africa. *Habitat International*, vol. 34, pp. 414–420.

Todes, A., and Harrison, P. 2001. The use of spatial frameworks in regional development in South Africa. *Regional Studies*, vol. 35 (1), pp. 65–72.

Todes, A., and Robinson, J. 2020. Re-directing developers: New models of rental housing development to re-shape the post-apartheid city? vol. 52 (2), pp. 297–317.

Turok, I. 2013. Securing the resurgence of African cities. *Local Economy*, Sage.

Turok, I. 2011. Deconstructing density: Strategic dilemmas confronting the post-apartheid city. *Cities (London, England)*, vol. 28 (5), pp. 470–477. doi:10.1016/j.cities.2010.10.003.

UN-Habitat. 2003. The challenge of slums. *UN-Habitat*, HS Number: HS/686?03E.

UN-Habitat. 2013. The state of planning in Africa. An overview. *African Planning Association*, UN-Habitat, HS Number: HS/010/14E.

UN-Habitat. 2015. Sustainable development goals. *United Nations*.

UN-Habitat. 2016. New urban agenda. *Habitat III*.

UN-Habitat. 2014. The state of African cities. *Re-Imagining Sustainable Urban Transitions*.

UN-Habitat. 2018. Leading change. *Delivering the New Urban Agenda Through Urban and Territorial Planning*.

United Nations. 2016. *Draft new urban agenda (habitat III)*. Retrieved from http://habitat3.org/the-new-urban-agenda/.

Watson, V. 2006. 'Best practice' planning and the de Soto thesis: Questioning the validity of universalism. *Planning Africa Conference*.

Watson, V. 2014. African urban fantasies: Dreams and nightmares. *Environment and Urbanization, International Institute for Environment and Development (IIED)*, vol. 26 (1), pp. 215–231.

Young, C. 2004. The end of the post-colonial state in Africa? Reflections on changing African political dynamics. *African Affairs*, vol. 103, pp. 23–49.

4

THE RESIDENTIAL, OFFICE, RETAIL AND INDUSTRIAL MARKETS IN WEST AFRICA

Emmanuel Kofi Gavu, Anthony Owusu-Ansah and Wilfred K. Anim-Odame

1. Introduction

The economic growth in Africa is underpinned by rapid urbanisation at 4.4% annually, the fastest rate globally. Sub-Saharan Africa, in particular, is experiencing a population growth which has more than doubled in the last 25 years, increasing from 470 million in 1990 to 1 billion in 2018. By the United Nations' projections, the sub-region will continue to record a growing population to 1.7 billion by 2040, representing 70%, and to 4 billion by 2100. In contrast, the population of the rest of the world is projected to increase by only 16% by 2040 (Anim-Odame 2016; Bahet al. 2018a, 2018b). This trend is expected to offer opportunities and challenges for socio-economic development at country levels. Such urbanisation and population growth rates, in the context of emerging economy, will stimulate growth in real estate markets across the continent and subsequently manifest in investment promotion.

Housing varies across many countries and regions, however a common feature among urban developing markets is an increase in demand for housing, which has resulted in increased prices out of the reach of the majority of the population (the poor and middle income earners) in need of same. Slum populations have continued to increase due to unaffordability of the bottom half of the income distribution. At the root of Africa's housing dilemma is rural-urban migration, currently averaged at 3.5% (the highest in the world). The rapid urbanisation rates, coupled with a lack of proper planning, has resulted in large housing deficits. Further to this, an accurate estimate of housing deficit is not readily available due to data asymmetries. Bah et al. (2018a, 2018b) argue that the mechanisms for creating a sound housing sector and delivering affordable housing to African households should be based on five pillars: urbanisation and infrastructure, land, construction, informal settlements and housing finance.

Within the urban context, the West Africa sub-region is characterised by deficiencies in land administration and land use policies. Cities are growing fast but with a lag of utility services and infrastructure in most peri-urban neighbourhoods. There is a general manifestation of new settlements that are informal and serviced by basic utilities such as pipe borne water and electricity. Governments in countries such as Ghana and Nigeria in the sub-region have

DOI: 10.1201/9780429279256-4

5

LAND TENURE SYSTEMS AND SUSTAINABLE LAND MANAGEMENT WITH SPECIAL REFERENCE TO THE KENYAN RURAL AREAS

Jennifer Murigu

Introduction

Tenure systems define who can hold and use land and land-based resources, for what length of time and under what conditions (Abdulai, 2006). These rules may be clear, well defined and documented or they could be ambiguous and open to misinterpretation and/or exploitation. When both formal and informal systems exist within a society, tenure rules can overlap leading in many instances to confusion, contradictions and insecurity. This has been the case in sub-Saharan Africa with as much as 90% of land falling under the informal system (Boone, 2017). It is estimated that about two thirds or 2.2 billion hectares of all cultivated land in sub-Saharan Africa is under customary tenure (Wily, 2011).

Kenya has multiple land tenure systems. Approximately 16% of the land in the country is owned privately, 24% publicly, while community land encompasses 60% of the land (Wily, 2018). The 60% of the community land is mainly within 21 of the 47 counties in the country.

This chapter focuses on the role land tenure systems play in sustainable land management with special reference to Kenyan rural areas. Different land tenure systems are evaluated holistically paying attention to the advantages and disadvantages of each system. Key issues related to land tenure systems and the effect each system has on sustainable land management are discussed.

The Concept of Land Tenure

The term land tenure is derived from the Latin word *tenere* which means "to hold" (Shipton, 2001). UN-Habitat (2008) defines land tenure as a set of rules, authorities, institutions and norms that govern the way people in a particular jurisdiction relate to land. Land tenure may be defined as the way land is held or owned by individuals and groups or the set of relationships legally or customarily existing amongst people with respect to land (FAO, 2002). The term defines the methods by which individuals or groups acquire, hold, transfer or transmit property rights in land.

Land tenure is an important part of the social, political and economic structures of any society. It is multi-dimensional, bringing into account not only economic but also social,

DOI: 10.1201/9780429279256-5

psychological, historical, technical, ecological, legal and political aspects that are often ignored but must be taken into consideration (Wanjala, 2000). Land tenure relationships may be structured, well-defined and enforceable in a formal judicial system or through customary structures in a community. Alternatively, they may be relatively poorly defined with ambiguities that are open to exploitation (Waiganjo and Ngugi, 2001).

Land tenure incorporates the concept of tenure security, which refers to enforceable claims on land, with the level of enforcement ranging from national laws to local village rules which are in turn supported by national regulatory frameworks (UN-Habitat, 2008). This reflects people's recognized ability to own, use, control and manage the land. Chigbu et al. (2017) define land security as the degree of clarity and certainty that a person's land rights will be recognized by law and by members of the community, and protected when there is a dispute. Security of tenure is the perceived assurance that one's rights to land will be recognized and respected by others as legitimate and protected in the event there is some dispute.

The Kenyan National Land Policy (2009) captures this in its definition of land tenure as terms and conditions under which right to land are acquired, retained, used, disposed or transmitted.

Resultant from secure tenure systems is secure land rights (USAID, 2010). They form a cornerstone for economic growth, economic development and an incentive for investment amongst other benefits, while insecure rights may lead to conflict, instability and the neglect of land. Clear and generally accepted regulations on the use of land aid in reducing the risk of overuse and degradation all of which will affect the environment negatively.

Land tenure constitutes various intersecting interests (FAO, 2002). These include:

- Overriding interests: Whereby a sovereign power for instance a government, be it the national or local government, a nation or community has the power to allocate or re-allocate land through expropriation. These are interests which although not protected in the land register are binding on any person who has an interest in registered land.
- Overlapping interests: This is whereby several parties are allocated different rights to the same parcel of land. They include leases, interests of persons in actual occupation other than beneficial interests under a trust, legal easements and profits à prendre.
- Complementary interests: This is whereby different parties share the same interest in the same parcel of land, for instance, where members of a community share common rights to grazing/pasture land.
- Competing interests: Whereby different parties contest the same interests in the same parcel, or instance, where two parties independently claim rights to the exclusive use of a parcel of agricultural land. Majority of land disputes arise from competing claims.

Land Tenure Classification

Land tenure systems determine who can use what resources, for how long and under what conditions. They outline the rights relating to the: use, control and transfer of the land; and the associated responsibilities and restrictions. According to Ogolla and Mugabe (1998), UN-Habitat (2008) and Kasimbazi (2017) land tenure is categorized as:

- **Private:** This constitutes the assignment of rights to a private party who may be an individual, a married couple, a group of people or a corporate body such as a commercial entity or non-profit organization. For example, within a community, individual families may have exclusive rights to residential or agricultural parcels of land with shared common areas like sports/play grounds, parks, walkways and driveways. Other members of the community will be excluded from using the private parcels of land without the consent of those who hold the rights.

- **Public:** Public land involves property rights being assigned to some authority for instance the national, regional or county government or a state. In most countries, forests and national parks fall under the mandate of the state, whether at a central or decentralized level of government.
- **Communal:** This is the right of members of a community to use a certain parcel of land. Each member has a right to use the holdings of the community. For example, members of a community may have the right to graze their livestock or farm on a common parcel of land.

In Kenya, the country's constitution under Articles 62, 63 and 64 categorizes land as:

- **Public Land:** The land is owned by the government for her own purpose. Categories of public land include: forest reserves, national parks, open water bodies, water catchment areas, land occupied by State organs, land transferred to the State, land to which no heir can be identified, land between high water mark and low water mark; and any land not classified as private or community land. Public land is held by the government in trust for the people of Kenya.

These lands are vested in the national government in trust for the people of Kenya and are to be administered on their behalf by the National Land Commission. The National Land Commission has powers to allocate or make grants of any estates, interests or rights in or over unalienated government land.

- **Private Land:** This is registered land held by any person(s) under freehold tenure, leasehold tenure and any other land declared private land under any Act of Parliament.
- **Community Land:** This is land lawfully registered in the name of group representatives, land lawfully transferred to a specific community and any land declared to be community land by an Act of Parliament. The Community Land Act (2016) provides for the recognition, protection and registration of community land. The Act repealed the Land (Group Representatives) Act and the Trust Lands Act which were Chapters 287 and 288 of the laws of Kenya respectively. Communities are identified on the basis of ethnicity, culture or any other defining characteristic. The community land may be held under any of the following tenure systems: customary, freehold, leasehold and any other system recognized under the Community Land Act (2016) or any other written law.

The Kenyan constitution (2010) acknowledges customary laws, only requiring them to be consistent with the constitution. The Land Act (2012) is more specific:

> "There shall be equal recognition and enforcement of land rights arising under all tenure systems and non-discrimination in ownership of and access to land under all tenure systems".

Land Tenure Systems in Kenya

Interests in land broadly fall under two categories: rights that are held through traditional African systems and rights derived and maintained through laws enacted by the colonial government that were later adapted and in some instance revised by the country's legislative body (Wanjala, 2000). The former is commonly referred to as customary law and the latter as the statutory law and expressed through various statutes/Acts of Parliament. Land tenure in most of African countries is either customary/traditional or statutory with roughly 90% of sub-Saharan Africa's land not dedicated to national parks and reserves and private land administered under customary tenure (Baldwin, 2017).

a. Customary Land Tenure System

Customary tenure is defined as the system of holding, utilizing and transacting in the land based on the traditions and customs of the people (Wanjala, 2000). Customary laws derive from the accepted practices and principles of the people. Customary land tenure is characterized by its largely unwritten nature where no land titles exist. It is based on local traditions, practices and norms, and is flexible, negotiable and location specific (Cotula et al., 2004). Its principles stem from rights established through first clearance of land or conquest. Customary systems are usually managed by a local or village chief, traditional/tribal authority or council of elders. Land under this tenure system is communally or jointly owned by particular groups of people.

Customary systems also differ in how members of various ethnic groups acquire, use, manage and transfer land. The Kenya Human Rights Commission (2018) estimates the number of ethnic groups in Kenya at 42. This number leads to multiple customary tenure systems, which vary based on the different agricultural practices, climatic conditions and cultural practices.

Prior to the introduction of colonialism, the predominant land tenure system was the customary tenure, where village elders and chiefs had the sole say on the allocation and use of land. At the structural level, the customary tenure was managed and protected by a social hierarchy organized in the form of an inverted pyramid with the tip representing the family unit, the middle the clan and lineage, and the base the community at large (Ogendo, 2000). These were decision-making levels designed to respond to issues regarding allocation, use and management of resources.

According to Ogendo (2000), most customary tenure systems exhibit a number of similar characteristics as follows:

- Individuals or groups by virtue of their membership in some social unit of production or political community have guaranteed rights of access to land and other natural resources. Individuals or families thus claim property rights by virtue of their affiliation to the group.
- Rights of control are vested in the political authority of the unit or community. This control is derived from sovereignty over the area in which the relevant resources are located. Control is for the purpose of guaranteeing access to the resources and is redistributive both spatially and intergenerationally. Its administrative component entails the power to allocate land and other resources within the group, regulate their use and defend them against outsiders.

- Rights analogous to private property accrue to individuals out of their investment of labour in harnessing, utilizing and maintaining the resource. For instance, the present cultivator of some piece of land has the greatest rights to it.

The regulatory mechanisms imposed by the political units such as exclusion of outsiders and seasonal variations in land use ensure sustainable resource utilization.

In Kenya, land under customary tenure occupies approximately 70% of the total area of the country (Wily, 2018). This mode of ownership in Kenya is currently governed by the Community Land Act of 2016. The Act recognizes, adjudicates and documents land rights for registration. Under the Act, land in the rural areas which is neither government land nor individually owned is vested in the county governments in trust for the residents living there. The Act empowers communities to make rules regulating the management and administration of their land. This takes cognizance of the fact that customary tenure systems are not static but evolve with time in response to changes in the economic, social, physical, ecological and political environment. Arguments for and against the customary tenure system are summarized in Table 5.1.

TABLE 5.1 Customary Tenure System: Advantages and Limitations

Advantages	*Limitations*
In difficult economic environments, customary land rights have proven for many poor people to be one of the few reliable assets over which they have secure control.	Persons who are not members of the community are usually excluded from acquiring land.
Every adult member of the community is a potential land owner.	Allotted land can be revoked by a new community leader(s). Also, people may lose rights when others ignore land tenure rules. Exploitation of unequal power relationships within communities, for example, may result in some members fencing off portions of communal land for their own exclusive use, thereby denying other members of the community access. Unmarried women may not always be allotted land.
Land ownership through this means is usually at no cost.	Permanent crop cultivation in most instance may not be accepted because land is allotted for only a farming season.
Communal land tenure system encourages large scale farming. Rural land holdings may be small, grazing land may be overstocked and degraded but the customary land right enables the family to build and maintain a home and engage in at least a limited range of agriculturally productive activities.	Mechanized farming may not be allowed and in cases where it is allowed, permission may have to be sought from the whole community.
Customary rights also ensure membership in a community of extended family. The members may be similarly poor but nonetheless a source of vital social capital and mutual assistance.	Land cannot be used as collateral for loan except when all the community approves it.

Source: Ogolla and Mugabe, 1998; Ogendo, 2000; UN-Habitat, 2008; USAID, 2010.

b. Statutory Land Tenure

Statutory systems of land tenure are based on written laws and regulations. These are Acts of centralized or decentralized government agencies and on judicial decisions. Land rights are allocated and confirmed through the issuance of titles or other forms of registration of ownership. The various forms of ownership under the statutory system are:

i. Freehold Tenure

This tenure confers the greatest interest in land called absolute right of ownership or possession of land for an indefinite period of time or in perpetuity. In Kenya, freehold land is governed by the Land Act and the Land Registration Act both of the year 2012. The Acts provide that the registration of a person(s) as the proprietor of the land vests in that person(s) the absolute ownership of that land together with all rights and privileges relating thereto. A freehold is the greatest interest a person can have on land as it gives the holder absolute ownership of the land for life. The land remains the permanent property of the freeholder and his/her successors, whether they obtain it as a gift, through sale or disposition on death.

Under Article 65 of the Kenyan Constitution (2010), non-citizens can only hold land for a period of 99 years under a leasehold tenure. This essentially means that a non-citizen cannot hold freehold property. The purpose of this provision has not been expressly stated in the Constitution or the land laws, but it may be seen as geared towards correcting historical injustices emanating from colonial rule where foreigners obtained large tracts of land and holding the same under freehold tenure in perpetuity. In most African countries, land individualization was demanded by the colonial settlers who required legal guarantee for the private ownership of land without which they were reluctant to invest (Ogendo, 2000). Kenya was no exception, and the privatization took place mainly in the high potential areas of the country (Wanjala, 2000).

With regard to the areas with lower agricultural potential, mostly arid and semi-arid parts of the country where the dominant land use is pastoralism, a different registration system was instituted in 1968 (Mbote et al., 2013). This was under the Land (Group Representatives) Act Chapter 287 of the Laws of Kenya. The Act together with the closely related Trust Lands Act Chapter 288 of the Laws of Kenya were repealed and consolidated under the Community Land Act of 2016. Under the Act, registration of group ranches was viewed as a compromise between individual ownership and the need for access to wider resources in arid and semi-arid areas. Under this system communal lands were divided into smaller units; ranches, which were then registered in the names of group representatives (three to 10 members) elected by the members of the group (Wanjala, 2000). Every member of the group has rights in the ownership of the group land in undivided shares. However, group ranches have with time been subdivided and converted to individual ownership.

Arguments for and against the freehold tenure system are summarized in Table 5.2.

ii. Leasehold Tenure

A leasehold is an interest in land for a definite term and it is usually subject to the payment of a fee or rent to the grantor (Ogendo, 2000). There are certain conditions which must be observed particularly relating to developments and usage. In Kenya, leases are granted by the

TABLE 5.2 Freehold Tenure System: Advantages and Limitations

Advantages	*Limitations*
Anyone is permitted to acquire any size of land based on their financial capability.	It is becoming increasingly expensive to acquire land through this means. Only the rich may acquire land in certain areas.
Land acquired through this system can be used for any means. Both permanent and arable crops can be cultivated. Animal husbandry is also very ideal.	
The owner has authority as to how the land is used, subject to the provisions of the law.	Land may be acquired in fragments. The small land holdings make it difficult to mechanize.
There is no discrimination as to tribe or area of origin in the process of acquiring land.	Prospective farmers may not have access to the required land.
Land can be used as collateral for loans.	Farm proceeds may not be adequate to service the loans.

Source: Ogolla and Mugabe, 1998; Ogendo, 2000; UN-Habitat, 2008; USAID, 2010.

TABLE 5.3 Leasehold Tenure System: Advantages and Limitations

Advantages	*Limitations*
A leasehold land based on the length of the lease allows the lessee to have maximum use of the land.	The leaseholder cannot develop the land beyond the lease agreement term.
It can be available for permanent crop cultivation depending on the lease period.	A lease agreement of less than 20 years may not be ideal for permanent crop cultivation.
There is no discrimination as to tribe or area of origin in the process of acquiring land.	The cost of the land may be prohibitive.
Land can be used by the owner as collateral for loans but based on the lease period.	Farm proceeds may not be adequate to service the loans. Also, based on the length of the lease or the unexpired term of the lease, plus other terms in the lease agreement, the land cannot serve as security for loans.

Source: Ogolla and Mugabe, 1996; Ogendo, 2000; UN-Habitat, 2008; USAID, 2010.

national or county government or by individuals holding freehold interests. The maximum term for government leases is 99 years (Land Act, 2012).

Once a government lease expires, the property automatically reverts back to the grantor but the leaseholder can apply for an extension of the lease. Section 13 of the Land Act (2012) provides that when the leasehold expires, the National Land Commission shall offer to the immediate past leasehold owner pre-emptive rights to be allocated the land provided that the lessee is a Kenyan citizen and that the land is not required by the national or county government.

In the case of individuals or organizations, the lease period is dependent on the agreement arrived at between the landlord and the tenant. Once the lease expires, the property automatically reverts back to the owner but the leaseholder may apply for an extension of the lease.

Arguments for and against the leasehold tenure system are summarized in Table 5.3.

Sustainable Land Management

The World Bank (2006) defines sustainable land management as a process of ensuring the requisite balance between environmental and ecological protection on the one hand and productivity of agriculture and forestry with respect to demographic growth and increasing pressure in land use on the other. Sustainable land management is necessary to meet the requirements of a growing population. Improper land management can lead to land degradation and a significant reduction in the productive capacity of the land.

The objective of sustainable land management (SLM) is to harmonize the complementary goals of providing environmental, economic, ecological and social opportunities for the benefit of present and future generations, while maintaining and enhancing the quality of the land. Sustainable land management aims at meeting changing human needs while ensuring long-term socioeconomic and ecological functions of the land.

Principles and Criteria for Sustainable Land Management

Sustainable land management (SLM) was defined at the Rio Earth Summit in 1992 as: "the use of land resources, including; soil, water, animals and plants, for the production of goods to meet changing human needs, while simultaneously ensuring the long-term productive potential of these resources and the maintenance of their environmental functions" (Sanz et al., 2017).

According to World Bank (2006, 2008) and Sanz et al. (2017), sustainable land management combines technologies, policies and activities aimed at integrating socioeconomic principles with environmental concerns, so as to simultaneously:

- Maintain and enhance production (productivity).
- Protect the potential of natural resources and prevent degradation of soil and water quality (protection).
- Be economically viable (economic viability).
- Be socially acceptable, and assure access to benefits resultant from improved land management (social acceptability/security of land rights/equity).

The evaluation of sustainability is therefore based on these objectives: productivity, equitable access to land, protection/security of land rights, economic viability, environmental viability and social acceptability.

The Kenyan constitution (2010) also provides that land in Kenya shall be held, used and managed in a manner that is equitable, efficient, productive and sustainable, and in accordance with the following principles:

(a) Equitable access to land; (b) security of land rights; (c) sustainable and productive management of land resources; (d) transparent and cost-effective administration of land; (e) sound conservation and protection of ecologically sensitive areas; (f) elimination of gender discrimination in law, customs and practices related to land and property in land; and (g) encouragement of communities to settle land disputes through recognized local community initiatives consistent with this Constitution.

titles were terraced; 15.2% of those without titles were terraced. If the first two are combined, as they are essentially in the same category only in the latter the titles are not physically with the owners, the percentage of terraced farms is 64.1%. This shows clearly the importance of obtaining title deeds (titling) on investment in sustainable land use.

Mwakubo (2002) also observed that the tenure system also determined the level of tree planting on the farms with an 8.4% tree planting level for the farms with titles, 7.6% for those in the process of obtaining titles and 0% for those with no titles. As Tiffen et al. (1996) argue, secure land tenure is important to farmers willingness to invest in land improvement, most particularly in long-term measures such as soil and water conservation.

6. Social Acceptability

Without landed property of any kind, poor communities and households are less able to absorb shocks associated with loss of livelihood in the formal and informal wage economies through which rural households augment their modest farming incomes (FAO, 2015). The land provides a means for the homesteads to invest in housing, livestock as well as crop farming (Lawry, 2014). The ability to benefit from these investments and to keep them within the families and communities intergenerationally would be diminished for the poor if land could only be secured through purchase.

Customary tenure is threatened by local and international interests that seize land for agricultural and other projects. These threats may be partially mitigated by treating customary land tenure arrangements as systems of economic, social and cultural rights analogous to those promoted by modern international rights conventions.

Conclusion

Traditional tenure systems often have evolved in response to livelihood and environmental sustainability needs, and changing these systems in the interest of making them more productive can be counter-productive. For instance, the Kenya government's attempt to enhance productivity in the pastoral areas by altering community-based land tenure systems failed to achieve this goal as customary land tenure systems tend to have different objectives other than those of achieving controlled and optimum productivity. Customary tenure systems are mainly geared towards equal and unlimited access to land by all the members of the community. They also have inherent ways of putting in place means of survival in harsh physical and economic environments. Hence the need for some balance between the private and customary tenure systems as each serves different purposes and under different circumstances.

The way the Kenyan government both at national and local levels governs land determines its tenure security and by extension its ability to sustainably harness its resources. The government holds the right to manage public land and to regulate customary land. There have been several initiatives by the government to initiate land reforms and in essence improve tenure security. Key to this is the promulgation of the 2010 constitution and the consolidation of the various land laws under the Land Act of 2012 with the aim of simplifying them and allowing for a less complicated and bureaucratic land management system.

However, legislation and policies are not an end in themselves. Effective, efficient and timely implementation is required.

References

Abdulai R. T. (2006); Is Land Title Registration the Answer to Insecure and Uncertain Property Rights in Sub-Saharan Africa? *RICS Research Paper Series* Vol. 6, April.

Abdulai A., Owusu V. & Goetz R. (2011); Land Tenure Differences and Investment in Land Improvement Measures: Theoretical and Empirical Analyses. https://econpapers.repec.org/article/eeedeveco/v_3a96_3ay_3a2011_3ai_3a1_3ap_3a66-78.htm

Baldwin R. (2017); Legitimate Land Tenure and Property Rights: Fostering Compliance and Development Outcomes. https://dai-global-developments.com/articles/customary-land-tenure.

Boone C. (2017); Sons of the Soil Conflict in Africa: Institutional Determinants of Ethnic Conflict Over Land. https://ideas.repec.org/a/eee/wdevel/v96y2017icp276-293.html.

Chigbu U. E., Schopf A., Vries W. T., Masum F., Mabikke S., Antonio D. & Espinoza J. (2017); Combining Land-Use Planning and Tenure Security: A Tenure Responsive Land-Use Planning Approach for Developing Countries. *Journal of Environmental Planning and Management* Vol. 60:9, 1622–1639, DOI: 10.1080/09640568.2016.1245655

Cotula L., Toulmin C. & Hesse C. (2004); Land Tenure and Administration in Africa: Lessons of Experience and Emerging Issues. www.hubrural.org/IMG/pdf/iied_lt_cotula.pdf.

Food and Agriculture Organization (2002); Land Tenure and Rural Development. https.//www.fao.org.

Food and Agriculture Organization (2015); The Economic Lives of Smallholder Farmers. An Analysis Based on Household Data from Nine Countries. www.fao.org/3/a-i5251e.pdf.

Government of Kenya (2009); National Land Policy, Government Printers, Nairobi, Kenya.

Government of Kenya (2010); Constitution of Kenya, Government Printers, Nairobi, Kenya.

Government of Kenya (2012a); Land Act, Government Printers, Nairobi, Kenya.

Government of Kenya (2012b); Land Registration Act, Government Printers, Nairobi, Kenya.

Government of Kenya (2016); Community Land Act, Government Printers, Nairobi, Kenya.

Kasimbazi E. (2017); Land Tenure and Rights for Improved Land Management and Sustainable Development, Global Land Outlook Working Paper, United Nations Convention to Combat Desertification (UNCCD).

Kenya Human Rights Commission (2018); Ethnicity and Politicization in Kenya. www.khrc.or.ke/publications/183-ethnicity-and-politicization-in-kenya/file.html.

Lawry S. (2014); Customary Land Tenure. https://dai-global-developments.com/articles.

Maxwell, D. & Wiebe, K. (1998); Land Tenure and Food Security: A Review of Concepts, Evidence and Methods, Research Paper No 129, University of Wisconsin, Madison.

Mbote P., Odote C., Musembi C. & Kamande W. (2013); *Ours by Right: Law, Politics and Realities of Community Property in Kenya*, Nairobi: Strathmore University Press.

Mwakubo S. (2002); Land Tenure and Farm Level Soil Conservation in Semi-arid Areas of Kenya. https://land.igad.int/index.php/documents-1/countries/kenya/rural-development-3/803-land-tenure-and-farm-level-soil-conservation-in-semi-arid-areas-kenya/file.

National Land Commission (2014); Advisory on Comprehensive Programme for Registration of Title in Land. www.landcommission.go.ke/media/erp/draft_advisory_comprehensive_ upload/program_booklet_for_registration_of_tittle_in_land.

Mohamed N. (2019); Conservation in Northern Kenya: Conflicts Over Community Land in the Pastoral Margins. https://pastres.org/2019/11/15/conservation-in-northern-kenya-conflicts-over-community-land-in-the-pastoral-margins/

Obunde O. P., Mbogo M. C., Kosura O. W. & Kamoni W. A. (2004); Effects of Land Tenure on Agricultural Productivity and the Environment. A Case Study of Suba and Laikipia Districts. IFPRI Eastern Africa Food Policy Network, Network Report 7. Kampala, Uganda.

Ogendo O. (2000); Through the Interstices of Procedure: Creating Democratic Space for Environmental Management. Conference paper for the East African Regional Workshop on Environmental Procedural Rights held in Entebbe, Uganda from 23rd to the 24th of November, 2000.

Ogolla B. & Mugabe J. (1996); *Land Tenure Systems: In Land We Trust*. Nairobi, Kenya: Initiative Publishers.

Payne G. & Durand-Lasserve A. (2012); Holding On: Security of Tenure – Types, Policies, Practices and Challenges. https://www.ohchr.org/Documents/Issues/Housing/SecurityTenure/Payne-Durand-Lasserve-BackgroundPaper-JAN2013.pdf

Place, F. and Otsuka, K. 2000. *The Role of Tenure in the Management of Trees at the Community Level: Theoretical and Empirical analysis from Uganda and Malawi*. CAPRi Working Paper No. 9. Washington, DC: International Food Policy Research Institute.

Sanz M. J., Vente J., Chotte J. L., Bernoux M., Kust G., Ruiz I., Almagro M., Alloza J.-A., Vallejo R., Castillo V., Hebel A. & Akhtar-Schuster M. (2017); Sustainable Land Management Contribution to successful Land-Based Climate change Adaptation and Mitigation. A Report of the Science-Policy Interface. United Nations Convention to Combat Desertification (UNCCD), Bonn, Germany.

Shimelles Tenaw K. M., Zahidul I. & Tuulikki P. (2009); Effects of Land Tenure and Property Rights on Agricultural Productivity in Ethiopia, Namibia and Bangladesh. www.ciesin.org/lw-kmn/slm/slm.html.

Shipton P. (2001); International Encyclopedia of the Social & Behavioral Sciences. https.//www.sciencedirect.com.

Tiffen M., Purcel R., Gichuki F, Gachene C, & Gatheru J. (1996); National Soil and Water Conservation Programme, SIDA Evaluation 96/25, Department for Natural Resources and the Environment, SIDA, Sweden.

UN-Habitat (2008); Secure Land Rights for All. United Nations Human Settlements Programme (UN-Habitat), Nairobi. https://dlc.dlib.indiana.edu/dlc/bitstream/handle/10535/6047/.

USAID (2010); Kenya-Land Tenure and Property Rights Profile 2. www.land-links.org.

Waiganjo C. & Ngugi P. (2001); The Effects of Existing Land Tenure Systems on Land Use in Kenya Today, International Conference on Spatial Information for Sustainable Development in Nairobi, Kenya from 2–5 October.

Wanjala S. (2000); *Essays on Land Law: The Reform Debate in Kenya*. Nairobi, Kenya: Faculty of Law, University of Nairobi.

Wily L. A. (2011); Customary Land Tenure in the Modern World Rights to Resources in Crisis: Reviewing the Fate of Customary Tenure in Africa. https://dlc.dlib.indiana.edu/dlc/bitstream/handle/10535/7713/customary%20land%20tenure%20in%20the%20modern%20world.pdf?sequence=1.

Wily L. A. (2018); The Community Land Act in Kenya Opportunities and Challenges for Communities. www.mdpi.com/2073-445X/7/1/12/htm.

World Bank (2006); *Sustainable Land Management. Challenges, Opportunities and Trade-Offs*. Washington, DC. https://en.wikipedia.org/wiki/Sustainable_land_management.

World Bank (2008); Agriculture for Development. https://openknowledge.worldbank.org/bitstream/handle/10986/5990/WDR%202008%20-%20English.pdf.

Veit P. (2011); Rise and Fall of Group Ranches in Kenya. file:///C:/Users/admin/Downloads/kenya-rise-and-fall-of-group-ranches.pdf.

6

THE IMPACT OF LAND TENURE ON REAL ESTATE MARKETS AND LAND-USE RIGHTS AND LAND REGISTRATION

Terzungwe T. Dugeri and Austin C. Otegbulu

Introduction

The Nigerian property market like those of most African countries has huge but unrealized investment potentials owing to its stage of development and opaqueness relative to her counterparts in Europe and Asia, and this has made it to suffer international investor apathy. This opacity is accounted for by many reasons, one of which relates to its weak structure status. The consistent poor rating of the land title registration processes across the states of Nigeria by The World Bank has been the one factor that has sustained this underperformance. Worried by Nigeria's poor showing on the World Bank's Ease of Doing Business Index in recent years relative to its peers, and with support of many international donor agencies, there have been sustained efforts on the part of federal and state governments to facilitate land title registration. It is argued in this work that the present state of the Nigerian property market is a direct result of the effect of the subsisting land tenure system on the functioning of the property market. This is demonstrated in the analysis of the Lagos and Abuja property markets in the context of land use rights and current efforts at achieving land registration.

This work is presented in four sections. The opening section which introduces the subject of study commences with a brief explanation of the terms and ends with an abstracted history that traces the evolution of the land tenure system in Nigeria. The second section is a review of related literature on land tenure systems and the role of property markets in the modern economy. The next section explains the methodology adopted for analysis of the selected markets while the last section provides explanations of the findings and its implication for policy actions needed to move these markets on their evolution path.

Land tenure is a term that describes and defines the system of land holding in a given society. The nature of land and its strategic significance to the immediate survival of man and in meeting the diverse needs of man has made it imperative for all societies to devise and encode practices for its holding. These land holding practices evolve and will normally derive from customs, beliefs, aspirations, and other historical events (Payne et al., 2015). These are usually expressed in the form of laws or regulations that specify the elements within the society who are charged with the execution or management of policy; those that are entitled and can

DOI: 10.1201/9780429279256-6

hold land; the ways by which land can be held; and the obligations and benefits of ownership (Feder and Feeney, 1991). In many instances these laws or codes would also define conditions for its expropriation from the land holding elements and the processes to be followed. *Land use rights* derive mainly from the subsisting land tenure system and are a set of rights pertaining to specific geo-spatial units needed to bring them to use either by the primary owning elements or by others deriving from the primary holders. While these are broadly set out by the extant land tenure system, land use rights are basically creations of contracts and serve their set out user objectives and provide for the management of the relationship between parties with concurrent rights over particular land space. *Land registration* is a process for the formalization of titles to landed property by demarcating and delineating specific parcels and documenting their ownership. It is usually achieved using geo-spatial technology for the capturing and archiving of unit-specific cadastral data.

Property market defines the complex processes in which property rights of diverse forms are exchanged among owners, users, developers, and investors directly or through other media in a given society or state. It is thus a generic term describing several submarkets which can variously be grouped spatially, by sector, or functionally such as the Lagos property market referring to a specific area; the commercial property market which comprises offices, shops and retail space, etc. by sector; the investment property market when functionally describing property primarily held for income; or mortgage housing submarket referring to the market for properties acquired through mortgage for housing (Dugeri, 2011).

A brief history of the evolution of land tenure system in Nigeria

Putting together a description of the evolution of the Nigerian land tenure system is like tracing the chords in a labyrinth of sort woven from different cultures and customs of the constituent peoples that comprise present day Nigeria. For the purposes of this work however, this history is viewed from three distinct phases namely: pre-colonial Nigeria, colonial era, and post-independence. This three-part segmentation is deliberate and in the perspective of the major occurrences that operated in each era to shape what has evolved into the subsisting land tenure system in Nigeria. This approach is also informed by the very limitation of having to account for each of the over 250 diverse people who make up Nigeria in a work of this nature and for enabling some distilling of the most common traditions, customs, and beliefs about land holding.

Land tenure among the different nationalities that comprise Nigeria as exists today was varied and dependent on each tribal group's custom and culture. In most of the northern part of Nigeria, the Hausa and other tribes upon being conquered by the Fulani jihadists were brought under Islamic rule and had their lands administered according to Islamic law. Prior to this and up to the time of the establishment of the Protectorate of Northern Nigeria by the British Colonialists, the unconquered parts of the north operated a customary tenure system. The other unconquered major tribes like among the Tiv of central Nigeria, land was held communally and devolved through eldest male heirs who held it in trust for their family members. Land among most tribes was not alienable to non-natives. Following Her Royal Majesty's troops' invasion and conquering of the ruling Fulani rulers that founded the Northern Protectorate in 1902, the Royal Niger Company who had earlier made acquisitions in the area in pursuit of trading expeditions along the River Niger, ceded their 'acquired lands' to the Crown. And all lands within the Northern Protectorate of Nigeria were brought under the Crown through a series of legislation imposed on the Protectorates.

In the southern part of Nigeria, the Lagos area and other surrounding south western communities comprising mainly of the Yoruba had a somewhat homogenous form of customary tenure system prior to the colonialization of Nigeria by the British. Among the Yoruba, land belonged to the community and was administered by the Obas working in council with subordinate chiefs and ward heads. Access to land was through membership of the community. Administration of lands was placed under the control and direction of the Baales (ward heads) as trustees for the people acting for the Oba. This practice was dislocated with the coming of Europeans who through treaties first made with King Dosunmu of Lagos and other chiefs had between 1863 and 1865 taken over lands within Lagos and its environs. The colonialists later on the basis of similar treaties with other chiefs in the hinterland gained a stranglehold on the lands of South Western Nigeria. On the south eastern flank of the Nigeria, a group of tribes comprising the majority Ibos, Efik, Ibibio, Ijaw, Itsekhiri, Urhobo, Bini, Ogoja, etc. operated a somewhat similar customary land tenure system. Among the Ibos, land is more than a physical resource and bears very strong spiritual relevance among the people. Land is held and administered communally with clan members deriving access through family heads. Land ordinarily devolves patri-lineally through sons as females are not entitled in their own names except through their male family heads as husbands. While transfers to strangers were prohibited, land could be exchanged among members of the clan with the consent and approval of family members. This system of land holding was preserved until contact was made with the British colonialists.

During the colonial era, upon amalgamation of the Southern and Northern Protectorates to create the young state of Nigeria in 1914 till the nation attained independence in 1960, the colonialists adopted several English laws. Commentators on the evolution of the Nigerian tenure system express a complete display of disdain by the colonialists for the customary tenure that subsisted in the country prior to their coming (Udoekanem et al., 2014). In the northern part of the country, having adopted the indirect rule system, the colonialists were bothered more with securing and excluding 'non-natives' from obtaining titles to land except where such lands were needed by the colonial government. They thus did not seek to radically disrupt the subsisting feudalistic tenure system imposed by the hitherto ruling Fulani Emirs in areas they had previously conquered.

In the southern part of Nigeria, earlier interaction and subsequent ceding of lands by Yoruba chiefs within the Lagos colony and later the southern western part of the Southern Protectorate saw to a gradual introduction of English land laws from transactions and treaties. Upon conquering the eastern flank of the south through a series of campaigns, the Ibo and other tribes also saw an imposition of English land laws though not in the manner implanted in the south west. The British colonialists encountered a highly decentralized non-monarchical system in contrast to that obtained with the north in the Southern Protectorate particularly in the south east. While the Yoruba of south west ran a different form of mornachy from the Northern Protectorate, the south eastern tribes operated a republican system of government. For this reason upon conquering peoples of southern Nigeria, the colonialists proceeded to recognize a few chiefs, created stools for others, and imposed additional powers on the selected chiefs. These powers were unknown within the customs and traditions of the south eastern communities (Gueye and Boahen, 1985). A colonial court and a police system were introduced, (as happened in all conquered territories), among the south eastern people to enforce the new imposed form of rulership which included powers to implement the colonialist's land laws.

In summary, land tenure and related literature points to the following laws as major legislations that emerged with the establishment of colonial administration up to the time of Nigeria's independence in 1960. In the southern part of Nigeria, the following legislations are identified to include: The Treaty of Cessation (1861) which ceded all powers of traditional title holders over land in the Lagos area to the British Crown (Nwabueze, 1974); The Land Proclamation Act 1900 and the Land and Native Rights Act of 1916 radically dislocated the fundamental basis of the subsisting customary land tenure system that vested land in the community and family and instituted the British Crown as the new owners whose representative, the governor, having the sole powers to grant rights of occupancy to both natives and non-natives, and whose consent is required to give effect to all transactions bothering lands with non-natives; the Niger Land Transfer Act also passed in 1916 operated to transfer all lands hitherto acquired by the Royal Niger Company to the British Crown and consolidated the colonial powers over land creating what would later become State Lands. The Native Lands Acquisition Act of 1917 was passed to formally regulate transactions on land by non-natives in the pre-colonial southern Nigeria. In the same year the Public Lands Acquisition Act 1917 which gave the new colonial masters powers to acquire land for creating the needed physical facilities and infrastructure to serve the government (see Udoekanem et al., 2014). In view of the subsisting land tenure system in northern Nigeria, the colonizing powers had little complications in securing control of the lands. The British upon conquering the Fulani emirs who controlled a significant portion of the vast lands in the north, with the exception of a few unconquered tribes of the north central belt, passed the Lands and Native Rights Act 1916 to effectively take over all lands in northern Nigeria. Thus, effectively the land tenure system which hitherto was communally held under customary tenure, by virtue of these laws and several other subsequent amendments made to these legislations brought about an end to the communal land ownership and passed same to the Crown.

Upon attaining Independence in 1960, the respective regions passed their own land tenure legislations which were patterned after what was inherited from the colonialists. For instance, in the north, The Northern Land Tenure Law 1962 was passed. The Northern Land Tenure Law 1962 copiously substituted the Crown as owner of lands in northern Nigeria with the state and retained all provisions which effectively put a nail on customary land ownership in northern Nigeria. This law though recognized the right of indigenous people to continued holding of their lands, it retained the policy of the Land and Native Rights Act of 1916 that reduced all that can be held into a right of occupancy to be granted for a specific term and on conditions imposed by the governor of Northern Nigeria as he deemed fit.

In a similar way issuing out of the 'form' of colonialization of the southern Nigeria area, the eastern and western regions had over a long period operated under the received English statutes and subsequent other colonial legislations on land (Nwabueze, 2009). It would appear that a land market had evolved and subsisted on account of formal recognition of individual transactions on the basis of received English laws which caused an erosion of the customary tenure principle of communal ownership. The introduction of Conveyancing according to English land laws and state grants to individuals introduced individualized land ownership – a radical departure from communal ownership of land. This unwittingly created a duality of tenure systems. Besides State Lands, out of which grants could be made to individuals, the extant practice upon colonialization recognized personal interests in land deriving from conveyance of lands purchased or acquired from customary tenure. The regional governments in southern Nigeria slightly modified and passed their respective land laws in the years leading up to Nigeria's independence (Nwocha, 2016). These laws did not change much from what was obtainable in the colonial era in southern Nigeria. It would appear that the 'new' tenure system imposed by colonialization which introduced individualization of land holding in these areas offered the elite and urban land owners' economic power (Nwoko, 2016). It was to be expected that as the instrumentality of a grant from the governor or the execution of a conveyance sequel to acquisition from the primary owners gained ascendancy and which interest was perceived to be secured with government recognition the succeeding government would have no incentive to radically alter it (Jegede, 1980). The situation at independence thus did not provide any impetus to consider changing the status quo in southern Nigeria (Oluwamotemi, 2010).

According to Oluwamotemi (2010), at independence the Northern Nigeria land tenure law offered easier access to land by government and investors, and it was relatively cumbersome and costly in the south. Compensation was payable in the north only for developments and not for land while in the south as the subsisting Public Lands Acquisition Act 1947 provided for, compensation was for both land and developments on land. The land market orientation of southern Nigerian regions contrasted sharply with what was obtained in the northern region. Complications arose in the wake of the new federal government's quest for land for development purposes particularly in the southern part of Nigeria and forced a search for a national land tenure system – a land tenure system that would make it easier for government and investors to access land for development needs and equally resolve the numerous issues that the southern regions' market driven tenure system caused (Udo, 1977). It is therefore not surprising that the federal military government then upon setting up several panels would adopt the salient provisions of the Land Tenure Law of Northern Nigeria of 1962 and shaped it into the Land Use Act of 1978 (Now Cap L5 Laws of the Federal Republic of Nigeria, 2004). The merit or otherwise of adopting those principles, considered outside the scope of this work, has remained a subject of debate till today. It is however the reason why most works on the evolution on land tenure in post-independence Nigeria consider the two laws – Northern Nigeria Land Tenure Law 1962 and the Land Use Act – as most significant in explaining land tenure evolution in Nigeria.

The promulgation of the Land Use Act in 1978 radically changed land tenure in Nigeria particularly in the southern part. The most notable provisions of the Act include the following:

i. Vesting of all lands comprised in the state in the governor of each state to hold in trust and benefit of the people;
ii. It provides that the only interest a person can hold in land is right of occupancy for a fixed term (either statutory right of occupancy for urban areas or a customary right of occupancy for non-urban areas) and on conditions stated in the certificate of grant;
iii. Requires the governor's consent for all transactions-alienation, mortgage, transfers etc.;
iv. Limits the size of land, age, and quantum of land an individual can hold;
v. Grants the governor powers to revoke/acquire any claims to land for overriding public interest or for default on conditions of grant;

vi. Provides for compensation in respect of land taken by government only for holder's/ occupier's unexhausted improvements or by resettlement;
vii. The Act reserves all matters relating to compensation claims to the Land Use & Allocation Committee (a statutory committee of the Act charged with advising the governor on land management policies – The Land Use Act Cap L5 (Laws of FRN, 2004) in Section 47 specifically removed the courts' jurisdiction from adjudication on same.

These are the main principles of land management of the Land Use Act that set it apart from previous land tenure legislation and have shaped land tenure in Nigeria for over 40 years. It is in the operation of these provisions that it is most noticeable in the various ways the property market is impacted. These principles which underlie all the provisions of this Act remain the thrust of this work and form the context within which the performance of the Nigerian property market is evaluated. To provide some working background, the next section is a review of related literature, followed by a description of the method employed for this study, and the final part discusses the observed impact of land tenure on the property market in Nigeria.

Related literature

Land tenure and the modern economy

Land tenure is an issue of primary importance to the economic evolution and development of any nation. Predisposed by a nation's political history, the people's custom, and religious orientations, land tenure systems vary across nations. The manner and processes by which a land tenure system affords citizens and investors access to land for diverse uses directly affects productivity in the economy. It is for this reason and other socio-economic motivations that nations take seriously the issue of making a workable tenure system. The significance of the impact of land tenure systems on productivity is most felt in agrarian societies or where the people rely on land for their livelihood and sustenance (de Soto, 2000).

In today's modern economy ruled by dominant world institutions and catalyzed by forces of economic globalization the need for national economies to prove competitive has never been better felt (Kievani et al., 2000). Best practices in the marketplace are translated into standards. These standards are continually evolving but originate from practices emanating from firm operations in the developed economies (Mattli & Buthe, 2003). The standards setting markets have evolved rapidly and are sustained by strong institutions (Dugeri, Gambo, & Ajayi, 2012) . It is therefore not surprising how the developing world particularly African nations with weak institutions whose economies are still emerging are faltering (Olukoshi, 2004). With sparks of successes in pockets of emerging nations, the competition is not losing steam. Over time, ratings of economies on specific aspects of the economy using indices set by the World Bank and other global institutions have emerged, and measured performance of economies are regularly reported. The most common of these is the Ease of Doing Business rating among others. The ratings have fed international investors swaying them in their investment decisions for emerging markets. Reliance on these ratings can be explained and justified in a significant way by the paucity of information on most emerging markets particularly for the under researched economies of the developing world like Nigeria and the other African property markets. The desire to partake in the international investment arena has imposed an urgent review of how base institutions like those governing land holding operate for effectiveness. It is therefore not surprising why Nigeria and her federating states struggle to earn a decent showing on the World Bank's Ease of Doing Business ratings among others. Thus it is considered an imperative to stay competitive and attractive to cross border investors. Built in the multi-factor metrics used to rate an economy, key property market indices namely employed include:

i. The procedures, time, and cost required to secure building approval permits;
ii. The procedures and costs to be incurred in registering property rights; and
iii. Contract enforceability.

While i) and iii) derive and form part of the operative environment of the property market, the ii) factor directly seeks to measure the ease with which a prospective investor can acquire and register property rights. The envisaged outcomes in measuring this metric can prove quite telling of the efficiency which with land administration agencies in a given society discharge their functions.

Fundamentally we can analyze the impact of land tenure on the property market by reviewing the land tenure law and evaluating how the tenure system impacts the efficiency of agencies and institutions or facilitates transactions on the market. From a land tenure perspective, it is discernible the types of interests that can be held in property, how these are defined and with this how the tenure system ultimately creates the form of property rights that are transacted in a given society. The operationalization of these rights is what plays out as procedures that prospective holders and investors must fulfil with the appropriate land administration agencies to give legal character to their property related entrepreneurial activity.

Method

This chapter relies on literature on land tenure and property market operations in Nigeria. This is supported with secondary data sourced from the World Bank's Doing Business Reports. The objective remains to show how the subsisting land tenure has impacted the Nigerian property market. The Lagos and Abuja property markets on account of their status have enjoyed higher research attention. The Lagos property market has been the most consistently covered in the Ease of Doing Business Surveys by the World Bank. The Abuja property market on the other hand, though not enjoying the same attention from the World Bank surveys, has been studied by different researchers recently. It is also one of the eight pilot states/units for the Systematic Land Titling and Registering project promoted by the Presidential Technical Committee on Land Reform on account of which it has benefitted from donor-sponsored studies. We have for the purpose of this work relied principally on the Ease of Doing Business Reports on Nigeria in the period 2011–2020, Ghebru and Okumo (2016), Ghebru et al. (2014), among others. While these works did not directly address the impact of land tenure on the property market, by appraising the efficiency of the land administration institutions, they capture root causes of the market's stifling. For instance, in measuring for property registration as an indicator for ease of doing business, the World Bank surveys tests for i) reliability of infrastructure, ii) transparency of information, iii) geographic coverage, iv) land dispute resolution, and v) equal access. The other two works directly appraise the structure and land administration systems and users' perception of the quality of service.

For this reason, descriptive analysis is employed in the synthesise for findings on account of the asymmetry of data collection metrics.

Findings

1. Constrained market environment

The property markets' operating environment envisaged by the principal law that governs land tenure in Nigeria – the Land Use Act (Cap L5 LFN, 2004) – intended that the government's land administration function dictate and direct operations of the property market in a manner that is not wholesomely restrictive. Ordinarily it is expected that government's role in a modern market economy as relate interventions in property market operations should be designed to be facilitative i.e. by ensuring that land administration systems are not just efficient but transparent. Direct control should only be subtly exercised, and on the occasion of targeted land reform policies aimed at addressing a serious tenure distortion. This is to be exercised only where the property market is indicted for imposing such distortions.

2. Over ambitious provisions of the Act

The events leading up to the promulgation of the Act and the urgency craved by the then federal military government of Nigeria did not permit a deep appreciation of the law's provision on property market operations. Section 1 of the Land Use Act vests all lands comprised in state in the governor to hold in trust and for the benefit of citizens. Having divested all persons of ownership of land, the law empowers the governor to grant rights of occupancy for a specific term and conditions to Nigerians and requires all holders of land to apply and obtain grants from the governor. This section narrowed the scope of ownership rights known to communities prior to the promulgation of this Act and fundamentally dislocated existing property market practices particularly in southern Nigeria. As at the date over 98% of rural lands in Nigeria are still in the effective control of communities whose traditional title holders as custodians must be consulted for acquisitions even by the government after over four decades of operation of this law.

The very requirement of issuance of a certificate of occupancy as evidence of grant by the government has operated to introduce and accentuate the character of informality into the property market. Given the objectives and provisions of the Land Use Act, all non-certificated holding or occupancy is deemed informal in the property market as financial institutions request for and recognize only certificated titles as collateral for loans and related transactions. It is rather odd that besides the cities of Abuja and Lagos, there is no urban settlement in Nigeria whose certificated or registered land holding covers up to 5% of the total area. This provision did take into account the capacity of the land administration system to process and issue certificates of occupancy. The processing of certificate of occupancy is one of the longest globally and has been one factor that has dragged Nigeria down on the Ease of Doing Business rankings. Nigeria between 2015 and 2019 scored 45.5, 47.4, 47.4, 51.1, 52.6 on registration of property indicator. This reflects very marginal improvement year on year for the five-year period.

It is equally worth stating that the government that imposed state ownership of land failed to provide the needed digital cadastre and associated logistics to give impetus to the implementation of the provisions of the Act.

3. *Partial or non-implementation of provisions of the Act*

State governors in Nigeria have equally failed to regularly constitute the statutory Land Use and Allocation Committees/Land Advisory Committees as prescribed by the Act. As a consequence, the needed velocity in property market transaction is stifled. While some governors have rightly delegated their consent duties to the relevant commissioners/directors charged with land matters, the very requirement of and need for consent to transact is repugnant to market operations and frustrates efforts at entrenching fluidity of market operations. It is belabouring and an unnecessary extension and retention of the colonialist laws that prohibited alienation to non-natives in the bid to have a tight control over land and landed resources. One of the main criticisms of the tenure law is the undue meddling in the marketplace by government through the instrumentality of the consent provisions. The colonialists sought to maintain their control over land where its sale was to foreigners. If this is the intention then the requirement for consent need not be a blanket one for all forms of alienation but made specific for this reason.

4. *Inappropriate and poorly developed bureaucratic machinery*

In the majority of the states, the land administration agency is poorly staffed, and staff are rarely provided with adequate logistics to undertake their work. The agencies are usually populated by a thin number of staff who sometimes do not possess the requisite professional qualifications and across most states are highly overwhelmed by the amount of file work. Operating manually and poorly provided with logistics, land administration agency staff saddled with land administration duties serve a disproportionately large number of clients transacting on various property rights and this creates a fertile environment for rent seeking. And the consequences are increased transaction costs occasioned by delays in virtually all processes, increased underhand payments, and reduced transacting velocity in the Nigerian property market. This manifests in the poor quality of the land administration function wherein it is difficult to entrench transparency. The Ease of Doing Business 2020 ranks Nigeria 183 out of 194 economies (World Bank, 2019).

5. *High costs of title registration and formalization fees*

There is a strong propensity for government acting through land administration agencies to impose high charges as fees for title registration and documentation of transactions, and in many instances processing of title registration is tied to payment of sundry taxes such as personal income tax and land use charge. These fees come in different forms categorized with the stages of processing and include the following in Lagos State:

i. Payment of search fee;
ii. Payment of charting fee;
iii. Payment of endorsement fee;
iv. Payment of ground rent (where property is covered of certificate of occupancy);
v. Payment of consent fees;
vi. Payment of capital gains tax;
vii. Payment of registration fee;

viii. Payment of stamp duty.

And tied to the processing/registering property title are these other sundry payments that should ordinarily not be a part of title registration fees but which add to the burden of complying with the nation's tenure law:

- Evidence or payment of income tax by showing a tax clearance certificate (TCC) which usually covers three years;
- Payment of land use charge; and
- Payment of Neighbourhood Improvement Charge (The World Bank, 2019).

6. Tenure induced property market opacity

A frenzy to improve on Nigeria's Doing Business Rating by state governments across the country has seen to the implementation of reforms for over a decade now (World Bank, 2014). The most relevant to property market is the digitalization of land administration systems in several states employing the tools of GIS. This is yet to yield meaningful results, and the expected gains may elude the property market if the quality of land administration is not improved on. The mere adoption of GIS tools and digitization as being implemented is proving too exorbitant for inclusiveness. For instance, in Kaduna State of Nigeria, one of the celebrated most improved in Doing Business Rating in 2018 who has adopted a GIS based land administration system, the focus on raking in revenue through the now digitized system has put the average cost of title registration in Kaduna at N250,000.00 ($694). Lagos State scored 9.0/30, just above the sub-Saharan average, in quality of land administration – a score significantly earned for poor transparency. Nigeria ranks 183 out of 194 economies on the registration of property rights index (World Bank, 2019). It is to be observed that while the market is notoriously information inefficient, the pervasive lack of transparency in the operationalization of the tenure law by land administration agencies in Nigeria has distorted the functions of professional groups serving the property market. Information is hoarded by both professionals and civil servants on key aspects of transactions, and it is difficult to gather information on market activities.

Recommendations

From the observed effects of the subsisting land tenure system on the Nigerian property market, it is vivid how the market has evolved slowly. It is proving difficult to pull through with land tenure reforms on account of the Land Use Act being embedded in the Nigerian Constitution, as amendments to the extant tenure law become a form of constitutional amendment. It will therefore be suggested that a property market route to ease its grip be considered in the following ways:

- Deliberate efforts must be made by government to de-clog those procedures bordering on consent by reserving such processes only where complexities are expected to arise in the title transfer and registration process; it is not enough to delegate and specify dates within which consents should be given;
- Constitute relevant statutory committees and enable these committees to function;

- Entrench good governance in land administration by improving on the processes and infusing transparency on information in government land information repository;
- Ensure that land administration agencies are adequately and professionally staffed and that needed logistical support for efficient operation is provided;
- Unbundle property rights registration from other taxes not directly related to the process to bring down the high cost of securing title to property through registration;
- Consider land registration as a service with long time benefits rather than engage the current practice of high charges upon digitization of land administration. This practice has denied current reforms in the land administration the touted inclusiveness and stunted rapid property rights registration;
- Lastly, fast-track ongoing reforms particularly in areas that improve the land administration systems through the adoption of a GIS based cadastre, removing the consent requirement clause, and recognition of customary tenure with a view to propping up the growth of the property market for the wellbeing of the economy.

References

de Soto, H. (2000). *The Mystery of Capital- Why Capitalism Triumphs in the West but Fails Everywhere Else.* London: Transworld Publishers.

Dugeri, T. T. (2011). An Evaluation of the Maturity of the Nigerian Property Market. *A PhD Thesis Submitted to the Department of Estate Management.* Lagos, Nigeria: University of Lagos.

Dugeri, T. T., Gambo, Y. l., & Ajayi, C. A. (2012). Internalising International Valuation Standards: Relevance and Applicability Issues in the Nigerian Context. *ATBU Journal of Environmenta Technology*, *5*(1), 100–116.

Feder, G., & Feeney, D. (1991). Land Tenure and Property: Theory and Implications for Development Policy. *The World Bank Economic Review*, *5*(1), 135–153.

FRN. (2004). The Land Use Act Cap L5 LFN. Abuja, Nigeria.

Ghebru, H., Edeh, H., Ali, D., Deininger, K., Okumo, A., & Woldeyohannes, S. (2014). Towards Security and DEmand for Land Tenure Regularisation in Nigeria: Empirical Evidence from Ondo and Kano States. *Nigeria Strategy Support Programme II, Working Paper No. 25.* Washington, DC: International Food Policy Research Insitute.

Ghebru, H., & Okumo, A. (2016). Land Adminstration Service Delivery and Its Challenges in Nigeria: A Case Study of Eight States. *Nigeria Strategy Support Program.* Washington, DC: International Food Policy Research Institute.

Gueye, M., & Boahen, A. A. (1985). African initiatives and Resistance in West Africa. A. A. Boahen & A. A. Boahen (Ed.), *Africa Under Colonial Domination 1880–1935* (Vol. 7, pp. 114–148). Berkeley, CA: United Nations Education & Scientific Organisation/Hieneman.

Jegede, M. I. (1980, January 17). *Land Law and Development -. Inauguaral Lecture Series.* Lagos, Nigeria: Lagos University Press.

Kievani, R., Parsa, A., & McGreal, S. (2000). Globalisation of Real Estate Markets and Urban Development in Central Europe. RICS Cutting Edge Conference. London.

Mattli, W., & Buthe, t. (2003). Setting Inernational Standards: Technological Rationality or Primacy of Power? *World Politics*, *56*(1), 1–42.

Nwabueze, B. O. (1974). *Nigerian Land Law.* Enugu: Nwamife Publishers Ltd.

Nwabueze, R. N. (2009). Alienations Under the Land Use Act and Declarations of Trust in Nigeria. *Journal of African Law*, *53*(1), 59–89. doi:10.1017/S0021855309000035.

Nwocha, M. E. (2016). Impact of the Land Use Act on Economic Development in the Country. *Admnistratio*, *2*, 117–128. Retrieved November 12, 2019, from www.ceeeol.com/search/article-detail?id=518028.

Nwoko, K. C. (2016). Land Ownersihp Versus Evelopment in the Era of GLobalisation: A Trajectory of Conflict and Wealth Accummulation in Southern Nigeria. *Journal of African Transformation*, 77–94. Retrieved October 16, 2019, from http://repository.uneca.org/bitstream/handle/10855/23204/b11562560.pdf?sequence=1.

Olukoshi, A. (2004). Property Rights, Investments, Opportunity and Growth: Africa in a Global Context. J. Quan, S. F. Tan, & C. Toulmin (Eds.), *Land in Africa: Market Asset or Secure Livelihood?* (pp. 25–34). London: IIED/Natural Resources Institute/Royal African Society.

Oluwamotemi, D. K. (2010, April 11–16). Land Acquisition, Compensation, an Resettlement in Developing Economies: Nigeria as a Case-Study. *FIG Conference 2010 Facing the Challenges-Building the Capacity*. SYdney. Retrieved October 17, 2019, from www.fig.net/resources/proceeedings/fig_proceedings/fig2010/papers/ts06e/ts06e_oluwamotemi_3616.pdf.

Payne, G., Mitchell, J., Kozumba, L., English, C., & Baldwin, R. (2015). *Legitimate Land Tenureand Property Rights: Fostering Compliance and DEvelopment Outcomes*. London: DAI.

Udo, R. K. (1977, March 17). *Land Policy for Effective Management of the National Economy-An Inaugural Lecture Delivered at the University of Ibadan*. Ibadan, Nigeria: University of Ibadan Press.

Udoekanem, N. B., Adoga, A. O., & Onwumere, V. O. (2014). Land Ownership in Nigeria: Historical Development, Current Issues, and Future Perspectives. *Journal of Environment and Earth Science*, *4*(21).

World Bank. (2014). *Doing Business in Nigeria 2014: Understanding Regulations for Small and Medium Sized Enterprises*. Washington, DC: The World Bank Group. Retrieved November 9, 2019, from www.doingbusiness.org.

World Bank. (2019). *Doing Business 2020: Nigeria*. Washington, DC: The Workd Bank Group. Retrieved November 10, 2019, from www.doingbusiness.org.

7

TOWARDS A HARMONIOUS LAND ADMINISTRATION INSTITUTIONAL SERVICE DELIVERY IN KENYA; A DECENTRALIZED APPROACH

Winnie Mwangi

1. Introduction and Background to Institutional Framework

Land has always been pivotal to the livelihood of Kenyans. People's capacity to construct a livelihood and overcome poverty in a country like Kenya is greatly influenced by their effective control over productive resources such as land. Borras and Franco (2007) note that in many developing nations like Kenya, land is still a significant contributor to the income of the populace since a major portion of their income comes from farming or farming related activities. This is despite far reaching livelihood diversification processes that have occurred in different places over time. In developing countries, lack of control over land is still strongly related to poverty and inequality. This centrality of land in people's livelihoods in Kenya has meant that land continues to enjoy keen attention of the central and local governments (Benjaminsen and Lund, 2003).

The governance systems and practices that have been used by most governments for purposes of land management and administration can best be described as fused, personalized and highly centralized resulting in poor delivery of land administration services. In Kenya, the background to this can be traced to the colonization by the British, where through various treaties and agreements most land that was 'not occupied' by the natives became Crown land. At the onset, various laws were enacted that saw the establishment of offices and institutions which were under the control of central government to administer and manage land. The East African (lands) Order-in-Council was enacted in 1901 which placed all public land under the control of Her Majesty. This later became the Crown Lands Ordinance of 1902 whose purpose was to regulate government control of the alienation of Crown land. This was later repealed to enact the Government Lands Act which conferred unto the commissioner of lands of the protectorate powers to alienate public lands in such terms and conditions as he might deem fit subject to any directions which the colonial secretary of state might give (Onalo, 1986). The Order in Council was later expanded and re-enacted in the form of Crown Lands Ordinance of 1902 whose purpose was to regulate the government control of the alienation of Crown land. Subsequently, the Department of Lands in 1903 was created to operationalize the 1902 Crown Lands Ordinance (Wanjala, 2000).

DOI: 10.1201/9780429279256-7

Legalizing of the colonial land administration continued with new legal provisions. In 1915, the Crown Lands Ordinance was repealed and Government Lands Act (G.L.A.) was enacted. This legislation gave the commissioner powers over land alienation either on leasehold or freehold terms.

In 1930, the Native Land Trust Ordinance was enacted and created the Trust Land Board to vest and manage the Native Reserves and the White Highlands. The Trust Land Board was to ensure that no land was acquired through private agreements, treaties or concessions with chiefs and traditional elders from the natives' lands by the settlers (Onalo, 1986).

In 1939, the Kenya (Native Areas) Order in Council established and vested the management of the native land to the Native Lands Trust Board (Onalo, 1986). This 1939 Order in Council could on attainment of independence become the Trust Land Act (Chapter 288) laws of Kenya which abolished the Native Land Trust Board and created the County Councils to manage trust land (Ibid). The County Councils (today called the county governments) hold the land in trust of the local people within their areas of jurisdiction up to date (Yahya and Swazuri, 2007).

The preceding discussion shows legal institutions and land administrative structures that were created by the colonial government to give the national and local government unfettered powers of decision making on land administration and management.

Bruce and Knox (2009) note that centralized land administration systems just like other sectoral programmes often birth discontent amongst the citizenry due to various reasons amongst them being poor services due to complacency by the institutions giving the services, lack of accountability on the part of the institutions since there are few checks and balances, and over-concentration of power to make decisions in a few offices often breeding corruption and fatigue amongst the citizens who have to keep travelling long distances to the same offices for even the most mundane service.

Fortunately, countries in transition have over the past two decades attempted to overhaul these centralized land administration systems and created lower hierarchy offices and structures that are close to the citizens, albeit with a lot of pressure for land reforms exerted by the civil societies and the citizenry. In Kenya, the clamour to reform land administration service delivery is traced to the clamour for a new constitution which started in earnest in the early '90s giving birth to the 2010 Constitution. Noticeably, politics involves governing the distribution of resources amongst competing priorities, and land is one such resource. In Kenya, Syagga (2006) points out that politics of equity has often been intertwined with politics of land sector reforms to accord everyone equity in land access and ownership. Thus, the process of constitutional reforms also saw the country formulate a land policy which came to be known as Sessional Paper No 3 of Land Policy, 2009. In the land policy, decentralization of land administration services was envisaged through creation of National Land Commission with its constituent offices in the various counties so as to bring services to the people.

Following closely came the Constitution 2010 in which principles guiding land management were laid out under Article 60 and the proposal to have the National Land Commission under Article 67. The constitution itself introduced a devolved system of governance where various processes of land administration are devolved. This means that the county governments in their respective departments now offer land administration services to the people.

2. The Need for Decentralization

The premise behind devolution is land reforms whose wider interpretation is not restricted to redistribution, restitution and resettlement but to reforming laws and institutions that govern the manner in which land is held, administered and managed. In Kenya, like in many countries in transition, decentralization has emerged as a valued political and economic tool whose key advantage is bringing services to the people in a participatory and accountable manner Bruce and Knox (2009). Proponents of decentralization justify it on grounds that it will provide for greater participation in public decision making which leads to an improvement in efficiency, equity, development and resource management (Agrawal and Ribot, 1999; Bruce and Knox, 2009). It is further argued that through this form of governance, decision making is brought closer to citizens increasing public sector accountability and effectiveness (Agrawal and Ribot, 1999). Centralization is associated with inefficient resource management, red tape, and distance from local needs and preferences (Marysse, 2004). Decentralization, which is usually empowered by legislation, has three levels: political, administrative and fiscal. Politically, there is governmental sharing of powers, while administratively in the different units of local government, formal decision making is primarily exercised by locally elected representative officials. Fiscally, the different units have their own budgets to allocate materials on a range of issues decided by law. The three facets of decentralization i.e. fiscal, administrative and political, should be implemented together and are what encapsulate decentralization of services (Marysse, 2004).

Decentralization envisages a certain level of autonomy in each of the institutions, though in land administration processes a loop system is most applicable since some outputs from some processes are used as inputs for the next process. In most instances, one department or level of government may not produce every land record, thus calling for a synchronized system for results and impact to be felt. In the case of Kenya, the National Land Commission was supposed to undertake certain mandates on behalf of national and county governments. However as alluded to by Boore et al. (2019), diverse actors in both administrative and representative institutions of the state, at both the national and county levels, were empowered as 'veto players' whose consent and cooperation was required to realize the reform mandate. Unfortunately, an analysis of land administration reform in eight Kenyan counties shows how veto players are able to slow or curtail the implementation of the new land laws. Therefore, the envisaged benefits have not been felt as expected.

The result has been institutional conflicts that keep on arising borne out of suspicions by the different agencies to the extent that the National Land Commission has had to seek for an advisory opinion from the Supreme Court for interpretation of its mandate. At the national and county government levels, consultation between the institutions that was to be one of the guiding principles in administration and management of land has been lacking. A score card report done by Land Development Governance Institute in 2015 showed that Kenyans were dissatisfied with the level of land administration services given by the NLC, national and county governments, due to extended timelines in service delivery and corruption (LDGI, 2015).

In view of this, certain concerns then arise which the author hopes to address in this chapter. What aspects of decentralization apply to land administration, should all facets and processes of land administration be decentralized to the lower government levels, and must there be conflicts between institutions mandated with land administration once the services are shared?

3. Decentralization, a Theoretical Concept

The term decentralization, according to Burns (2007), is a multi-faceted generic concept that includes concepts like devolution, deconcentration, delegation and delocalization. It is important to note that any decentralization policy will most likely comprise doses of each and thus the need to understand what each concept involves. This enables us to draw a nexus between decentralization and land administration later. Kauzya (2007) and Burns (2007) view these concepts of decentralization as follows:

Deconcentration – This is the process of administrative decentralization whereby the central government designs a structure that enables its agents to work close to the local people in field units/agencies of central government.

Delegation – This involves the transfer of responsibilities from central government to semi-autonomous bodies that are directly accountable to the central government. Government agencies whether centrally located or at the locations in the periphery regions are examples of delegated responsibilities.

Devolution – It is the process of transferring decision making and implementation powers, functions, responsibilities and resources to legally constituted local governments. In such cases, the national government either shares some powers and functions with the local authorities or relinquishes certain key functions to the local authorities which presumably are closer to the people.

Delocalization – The spatial distribution of central government socio-economic development facilities and activities such as schools, hospitals amongst others in peripheral regions is what is referred to as delocalization.

Since its entrance in the policy agenda arena in the 1990s, decentralization has been a critical component of any governance reform strategy. The main objectives in these reforms have been geared towards transfer of management, functions and offices from a central place (central government) to peripheral regions (various local institutions). However, Bruce and Knox (2009:1361) argue that "decentralization is not simply a matter of what powers will be devolved to local authorities, but of *actual legitimacy* and *power* of the various institutional actors at the local level." According to Kauzya (2007), a successful decentralization reform process should embody both vertical and horizontal decentralization where vertical decentralization involves transfer of power and authority from central to local government. Horizontal decentralization on the other hand empowers the local communities and enables them to receive and utilize the powers that are transferred to them especially in problem analysis, priority setting and planning. The local communities are expected to constantly demand accountability from their local and national leadership or any governance actor at the local level (Ibid).

3.1 Dimensions of Decentralization

Whatever mode of decentralization is adopted, there are various 'indicators' that determine what level of success the decentralization programme will achieve. These parameters, also called dimensions, are the drivers or motivators of the decentralization process. Governments often use any or all three dimensions depending on what they hope to achieve. These are as follows:

Fiscal Decentralization – In its general use the term fiscal is mainly associated with finances; however, as used here it "comprises the assignment of own-source revenues to sub-national governments." Analysts hold that, due to the almost universal deficiency of local governments/bodies/institutions own source revenues relative to the assigned expenditure requirements there is need for inter-governmental transfer to alleviate the regional disparities (Smoke, 2003).

This dimension is critical if decentralization is to have the desired impacts, and according to Ahmad et al. (2005) it is defined by the following four components: (i) allocation of expenditure responsibilities by central and local tiers of government, (ii) assignment taxes by government tiers, (iii) the design of an intergovernmental grants system and (iv) the budgeting and monitoring of fiscal flows between different government tiers.

Institutional Decentralization – It is concerned with the critical institutional architecture on which decentralization is built. This in essence refers to the administrative bodies, systems and mechanisms, both local and intergovernmental, which help to manage and support decentralization. As Smoke (2003:10) also notes it also includes "mechanisms that link formal government bodies to other key local actors" such as the traditional authorities, private sector and the civil society.

Decentralization, however, without appropriately designed and implementable structures and processes as well as adequate local capacity to manage the political and financial functions of the local governments tends to fail (Burns, 2007; Kauzya, 2007; Ahmad et al., 2005).

Political Decentralization – This dimension implies the need for an inclusive local political process. Kauzya (2007:4) describes political decentralization as both "voice and the *vote*, and essentially leads to an enhancement of the influence of local people on the decisions that concern them." In analyzing this form of decentralization, it is important to analyze and assess "the processes and mechanisms through which the consensus is engineered and reached" (Kauzya, 2007:6).

3.2 Decentralization: The Need for a Cautious Approach

Kauzya (2007) observes that opting for decentralization does not necessarily guarantee that the population is invited to participate in decision making which is the goal of a good policy of decentralization and good governance. The argument brought fore by Kauzya (2007) is that there is a need to establish frameworks through which the population is not only assured of participation, but also inclusion in all decision-making processes with redress mechanisms in place.

Bruce and Knox (2009:1361) note that key for the success of decentralization is a governance system that recognizes that "local institutions can better discern and are more likely to respond to local needs and aspirations." As such, governments should not embrace decentralization as a 'catch-phrase' without the actual commitment that it will result in common good. This is because the instruments of determining if decentralization will succeed or not still lie with the state, and the state holds the benefits that will result from decentralization as the goal in its hands. It is this kind of thinking that has scholars like Smoke (2003:9) supporting decentralization on the premise that it will "improve efficiency, improve equity, improve governance and improve development leading to poverty reduction." However, as Bruce and Knox (2009) note, the mere presence of the decentralization policy will not necessarily lead to the aforementioned desired goals which to a large extent will depend on how it is structured and implemented.

As argued earlier, as much as decentralization is important, it is equally important that we are not blinded by its potential goals because the transfer of decision-making power to the local level is not necessarily a panacea, as the local institutions are not necessarily more democratic, more efficient or less corrupt than the central government (Mwangi, 2012). In addition to this, it is important to note that not all functions can best be carried out at the local level and it is a fallacy to assume that local authorities/institutions have better access to information and are more accountable to local populations on the mere fact that they are close to the people at the grassroot levels.

On this, Bruce and Knox (2009) assert that even where some tasks may best be accomplished at the local level, the results may be limited due to challenges of personnel and technology to actualize them. Besides, any form of decentralized service must be accompanied by some or all the elements of decentralization for them to succeed. Notably there has to be financial capacity to support the programmes, institutional set-ups must be put in place so as to avoid vacuums or duplication, and all this must be crowned by political goodwill to have the programmes succeed.

From the preceding it is apparent that, as it concerns decentralization, it is important to adopt a cautious approach as opposed to a naïve, exaggerated positive outlook. In supporting this approach, this chapter is of the view that any envisaged decentralization policy should aim at coming up with effective and appropriate strategies (Bruce and Knox, 2009:1361).

4. Land Administration Decentralization

As is the case in many countries, discussion surrounding the land sector often takes centre stage in development policy debates. This is because land is the basis upon which any form of development takes place, and it is the very source of livelihood to the citizens. Management and administration of land resources is, therefore, at the heart of both citizenry and government. Delivery of secure land rights has been found to precede sustainable development since people become more accountable for the land resources at their disposal (UNECE, 1996). As such structures, institutions, finances and human resources always have to accompany successful land administration and management programmes. FAO (2007) echoes the critical role of the state in administering land resources, often because the state is the radical title holder in most countries. This argument has often been misused by state bureaucrats that land administration is a state affair, hence centralizing almost all the functions. However, as Mwangi and Nyika (2018) show, other gatekeepers are key to the success of land administration programmes. This has been supported by the proponents of land reforms (Barraclough, 1999; Adam, 2000) to show that the state cannot reform itself and must therefore accommodate other players in land administration reform processes, often through decentralizing some of the key land administration functions.

Land management encompasses "all the activities associated with the management of land as a resource, from both an environmental and economic perspective, towards sustainable development" (Burns, 2007:4). As a process it underpins "the distribution and management" of land resources which is one of society's key asset (Ibid). The main aim of land management supported by an efficient land administration system is to "deliver efficient land markets and effective management of the use of land in support of economic, social and environmental sustainability" (Deininger and Enemark, 2010:xviii).

Land administration on the other hand refers to the processes of determining, recording and disseminating information about tenure, value and use of land when implementing land management policies (Dale and Mc Laughlin, 1999; UNECE, 1996). Thus, land administration is an integral part of land management alongside land information system and land legal framework. Land administration is a vehicle in which land management is delivered and it operates with an institutional and regulatory framework established by land policy and the legal, social and environmental background of a particular jurisdiction (Deininger and Enemark, 2010).

A typical land administration system is comprised of an information system that defines rights and/or information, and spatial records that define the extent over which these rights and/or information can be exercised. Mwangi and Nyika (2018) argue that though not limited to the following aspects, a typical land administration system should cover the following areas:

i. Management of public land
ii. Recording and registration of private rights in land
iii. Recording, registration and publicizing of the grants or transfers of those rights in land through e.g. sale, subdivision, consolidation
iv. Management of the fiscal aspects related to rights in land, including land tax, historical sales data, valuation for a range of purposes, including assessment of fees and taxes and compensation for state acquisition
v. Control of the use of land, including land use zoning and support for the development application/approval processes

Land rights registration, surveying and mapping of boundaries make up the core land administration functions which are primarily the mandates of the state/public sector. This, therefore, makes land administration to be mainly a concern of the state, as it is the state that develops, maintains and safeguards the public infrastructure needed to support and protect land rights (Ibid). Thus, land administration in most countries is in the hands of state agencies and hence highly centralized, more so in the developing countries. In countries such as Kenya, which have a colonial legacy, there exists a dual land administration system because of the mix between the modern (read colonially adopted), and customary land administration systems. The recognition of customary and communal land tenure in land administration systems, the world over, is varied with some having explicit recognition as is the case in the Philippines (Kummer, 1992) and Bolivia. However, it is important to note that even in these contexts the administrative systems operate in a very complex and conflicting policy, legal and institutional environment resulting in limited tenure security particularly in the rural areas since administration and governance structures are centralized to either the central or local governments with limited delocalization. Other jurisdictions such as Uganda and

Mozambique operate under a "unified legal system based on customary law." In the extreme cases, this tenure type is not formally recognized, as is the case in Thailand (Burns, 2007:6). This explains the huge role that central government has in directing land administration in many jurisdictions.

4.1 Decentralization of Authority Over Land Administration

The institutional structure that supports land administration services is key in determining the efficiency of the land administration system. Land administration systems may vary from single centralized systems as is the case in Australia to decentralized systems like in Thailand, where the title register is split among 76 provinces and 272 branch provincial office, each maintaining the land administration within its jurisdiction (Burns, 2007). In Africa (Bruce and Knox, 2009), decentralization of authority over land may be traced back to the colonial period when colonial masters entered treaties with traditional leaders on how land was to be owned and distributed between themselves and the native local communities. On gaining independence, the post-colonial governments adapted strategies that ranged from strategic cooperation to outright efforts to dump traditional authorities. One way of achieving the latter strategy was through centralization of land control (Bruce and Knox, 2009). However, even in situations where land administration was centralized, the traditional authorities remained influential and important to the local people making their role in any decentralization of land management and administration policy central and hard to ignore. In Kenya, for instance, the now repealed Trust Land Act Cap 288 gave local authorities the mandate to administer and manage trust land on behalf of communities within their jurisdiction. The understanding, therefore, was that all land transactions falling under the local authorities had to be authorized by the county councils before being ratified by the central government (Onalo, 1986). We can, therefore, say that in this context certain aspects of land administration were decentralized while others remained centralized.

As detailed next, Upton (2009) best captures the advantages and disadvantages of a decentralized land management and administration system. One of the key benefits identified is the reduction of dominant state policies and practices which often muzzle the voices of the majority at the grassroots. The resources' rights of the local users are also prompted albeit through the support of customary arrangements which is better embraced by the locals. Active participation facilitated by local representatives leads to buy-in and ownership of land resources management programmes leading to sustainable land management (Ibid). However, issues of equity, inclusivity and contested definitions of customary rights need to be factored in for optimum benefits to be realized.

Of importance to this chapter as it concerns the decentralization of land administration services is the distinguishing of reforms that simply put in place local representatives of the central government i.e. *deconcentration*, and those that confer a degree of decision-making powers upon the local institutions/bodies i.e. *devolution*. Though both seek to improve the efficiency of services delivery at the local level, one is merely an extension of the central government at the local level whilst the other empowers the local institutions. This differentiation is important because deconcentration of authority over land is "an attempt by the state to realize more of its legal authority over land at the expense of local institutions" whilst devolution on the other hand "enhances local control over land by either traditional or local civil authorities, depending upon whom authority is devolved to" (Bruce and Knox, 2009:1362).

4.2 Institutionalizing Devolution Towards Land Administration Governance

In recent times, there has been a move towards greater devolution based on arguments like "devolution creates possibilities for more autonomous decision making in response to local needs circumstances, more room for citizen participation" and enhancement of downward accountability (Hillhorst, 2010:2). As it concerns land governance, institutional arrangements have been deemed an important and critical element in any land administration decentralization reform, especially when viewed in the context of the roles of central government, local authorities and community or customary authorities (Burns, 2007).

Some of the key local institutional actors in the decentralization of authority over land include:

a) Local Government

Where they exist, local governments are included in the formal implementation structure of land administration. In Ethiopia, for instance, they are in charge of land registration and also support for land administration services (Hillhorst, 2010). Due to the central role they play in society, as it concerns local governments, a combination of devolution and decentralization of land administration services is usually undertaken. In addition to securing the local tenure rights these institutional bodies may also play several other roles such as boundary surveying and mapping (Ibid).

b) Community Level Committees

These are committees established at the village or local government level to aid in implementation of land policies. Members of these committees may be elected, proposed by the community or appointed. According to Hillhorst (2010:8) these institutional structures "facilitate; the creation of rights inventories, the marking of boundaries, registration of land, recording transactions and deeds safeguarding"; an example of these committees is the land administration committee in Ethiopia at the *kebelle* level. The status of these committees and their relationship with the other land administration is varied, and thus attention needs to be paid on how they execute their mandate, their composition and their levels of accountability so as to achieve good (Ibid).

c) Customary Authorities

Though traditional authorities have played a critical role in land administration services provision, formal involvement of these authorities in land administration is rare. In addition to this is the fact that the legitimacy of customary leaders is contextually varied and is majorly impacted by the commodification of land. A common criticism against this institutional arrangement is that it does not adhere to (formal) democratic procedures, entrenching the position of local elites through exclusion of certain groups. As a strategy, this institutional arrangement should be adopted with a sense of caution and realism (Hillhorst, 2010).

5. Review of Decentralization of Land Administration in Kenya

In a study undertaken by the author between 2011 and 2012 in Narok County, the general understanding by citizens was that devolution is where information on land administration, the services, the relevant staff and the offices are brought close to the people. The expectation was that members of the community are involved in the decision making with regard to management of the local land and land-based resources, collection of revenue, planning and allocation of revenue to various uses within the community (Mwangi, 2012). Devolution was also seen to offer a chance for community participation, representation and inclusion in the appointment of officers in the local land institutions.

In the Narok County context, the concern of the people regarding land administration was around the following areas (Ibid):

i. Dispute resolution over conflicts on title land ownership
ii. Registration of titles and issuance with titles
iii. Transfer of titles and lease/sale agreements
iv. Land mutations and sub-divisions services
v. Adjudication and ascertainment of boundaries and ownership of land
vi. Cadastral maps (Registry Index Map) production

These services fit within the bigger spectrum of the core areas of land administration and are services often required by most people in the rural and peri-urban areas of Kenya. Other common services include approval of development applications for land subdivisions, extenuation of leases, change of user and building plans approvals. All these fall within the wider areas of land registration, adjudication, survey valuation, planning and allocation. In the context of this chapter, the author uses the following indicators in an attempt to review the performance of decentralization of land administration in Kenya, namely, functions and institutions, human capacity, funding, land information systems and legal framework.

5.1 Functions and Institutions

An examination into how land administration functions have been decentralized since the new constitutional dispensation paints a dull picture. There has been a disconnect between the institutions rendering land administration, particularly between the Ministry of Lands and the National Land Commission (Mwangi, 2012). The Land Policy 2009 envisaged that there will be County Land Management Boards (CLMBs), arms of the National Land Commission, that were to be devolved to the lowest administrative units; practically this would be at the sub-county levels. CLMBs were to be constituted by the county governments in consultation with the National Land Commission. However, as per Boore et al. (2019), about 10 of the 47 counties have not constituted these boards up to date. In its 2014/2015 annual report, NLC reported that some county governments have been uncooperative in constituting the boards while in other counties the officers lack adequate office space or shared offices making the execution of their duties difficult.

Furthermore, the legal framework within which the boards were to operate was not done, since the national government which formulates policy did not repeal the Land Control Act, Cap 302. This legislation operationalizes the land control boards that have been in existence since independence to regulate transactions on agricultural land and all freehold lands.

The implication of this scenario is that there cannot be two parallel boards and this renders any CLMBs formed toothless as long as the older land control boards have not been disbanded. In the absence of CLMBs it also means that NLC has only devolved offices to the county levels but the mandates of the officers are confined to limited roles as the other land transactions are either processed by the local land control boards or the county governments.

In examining the decentralization from a function point of view, besides physical planning which is a devolved function, the other core functions such as surveying and titling are directed by the national government with peripheral roles played by the county governments. The mandates of the National Land Commission, even though well intended, have continued to be clipped with the Land Laws Amendment Act 2016 and further statute amendments of 2018 to the Land Act. Thus, the key roles of the NLC have been management of public land, compulsory acquisition and addressing issues of historical injustices. The amendments to the Land Act saw the NLC lose the role of determination and collection of rents, meaning that statutory valuations are functions of the national and county governments.

5.2 Human Capacity

Human capacity that is capable and well-motivated plays a key role in delivery of successful land administration (FAO, 2007). Whereas there has not been a critical shortage of officers at the decentralized land offices, funds for salaries and capacity building as well as training programs have been limited. Devolution in Kenya came with huge financial ramifications since county governments recruited their own staff instead of absolving those who were employed by the national government. In some counties there are instances where we have county, NLC and national government land officers all doing the same functions thus resulting in duplication of roles while at the same time leaving some officers idle. This makes the labour force bloated and expensive to maintain. This scenario has often caused confusion over who is supposed to undertake what roles, often delaying services sought by the citizens. Besides these overlaps, both levels of government have not held meaningful capacity building amongst their staff and local communities to clarify what services are rendered by what office.

5.3 Funding

Funding is one of the critical aspects of decentralization. In Kenya, all revenue collected by the various arms of government is administered by the national treasury. As such, county governments and National Land Commission like all other agencies must come up with programmes and budgetary estimates of their activities with the current year and forward the same to treasury. Complaints however abound between the national government on one hand and the county government and National Land Commission on the other hand. This causes delays in disbursements of funds and in reduced budgets, thus limiting expenditure of the land administration and management programmes that these agencies might have. The Ministry of Lands has also not been able to fully roll out land registration and adjudication programmes due to limited financial capacity. This is echoed by NLC in its 2014/2015 annual report reported that NLC

> continued to face financial challenges owing to low budgetary allocation by government. This has constrained the Commission in undertaking planned activities the Commission had a deficit of Kshs 162.5 million during the period under review in its total allocated budget for the financial year. As a result the Commission was unable to

> undertake optimally some of its key activities because of inadequate transport, lack of visibility and awareness programmes, inadequate staff among others.
>
> *(NLC, 2016:56)*

5.4 Land Information System

Availability of up to date and verifiable land information is vital in decentralization of land administration. A modern land information system backed by modern information computer technology (ICT) allows for easier access and sharing of land information between all the land agencies (Dale and Mc Laughlin, 1988). In a centralized system, one or a few state agencies not only generate but keep a repository of the land records. Shifting from a centralized system to decentralized system means that some land records, especially when they are in manual form, will have to be transferred to the new agencies while at the same time some records will be shared between the agencies. In Kenya, the birth of National Land Commission which is the key custodian of public land meant that files, survey records and other support documents would have to be moved from the Ministry of Lands to NLC or mechanisms put in place for NLC to access the records when necessary. However, there have been transitional challenges arising from conflicts between the Ministry of Lands and NLC regarding the general mandates of each agency and the records that each is supposed to generate and keep (Mwangi, 2012; LDGI, 2015). The NLC observed:

> The transitional change is at a slow pace especially when it comes to accessing files which are available in the Lands Registries thus the Commission has managed to access a few documents due to challenges in accessing supporting attributed data and some digitized parcels that do not have ownership and other related information.
>
> *(NLC, 2016:54)*

5.5 Legal Framework

Decentralizing land administration requires a clear legal framework devoid of confusion and/or misinterpretation (LDGI, 2015; FAO, 2007). The Constitution 2010 introduced a devolved system of governance and as per Article 67 some functions were given to the National Land Commission while Fourth Schedule devolved physical and land use planning to the county governments. The enactment of Urban and Cities Act, 2011, Environment and Land Court 2012, Land Act and Land Registration Act 2012 and amendment of Physical Planning Act to Physical Planning and Land Use Act, 2019 have all been towards supporting a decentralized framework as far as land administration in Kenya is concerned. However, legal framework must be backed by political and institutional goodwill for it to deliver the envisaged benefits that come with change of legislation. This has not been demonstrated to be the case. Public spats particularly between the NLC and the Ministry of Lands have been a common occurrence with the latter reading mischief in the various amendments to the substantive legislation, often undertaken without the involvement of the NLC. Once again in its annual report 2014/205 NLC noted:

> Despite very clear policy and constitutional backing, the Commission has spent close to three years trying to get a key partner in the land sector to appreciate the Commission's

critical space in land administration and management. The deliberate misinterpretation of roles has caused confusion as to who was responsible for which mandate. This challenge has persisted even after the Supreme Court delivered an advisory in December 2016. In addition to these challenges, the Commission has noted increased reluctance by state agencies to observe the seven principles of land management as anchored in the National Land Policy.

(NLC, 2016:54)

The preceding analysis is an attempt to examine the extent to which decentralization of land administration has taken place in Kenya. As observed, certain bottlenecks need to be cleared but more importantly, there should be genuine commitment from all the stakeholders responsible for delivering access to secure land rights in a devolved system.

6. Conclusions and Way Forward

In making a case for effective land administration decentralization, the author has shown that though land is a critical resource in Kenya, its management and administration has remained centralized with only a very limited degree of decentralization. It has been noted that proposed legislations as far as it concerns true devolution of land administration are still wanting.

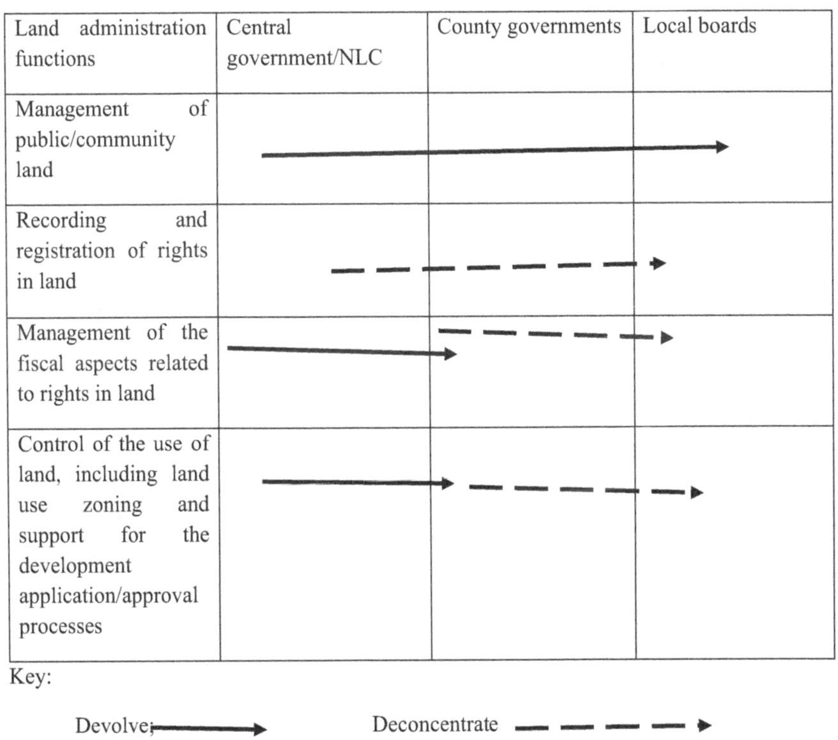

FIGURE 7.1 Proposed Land Management/Administration Decentralization Relationship Model

Source: Mwangi, 2012

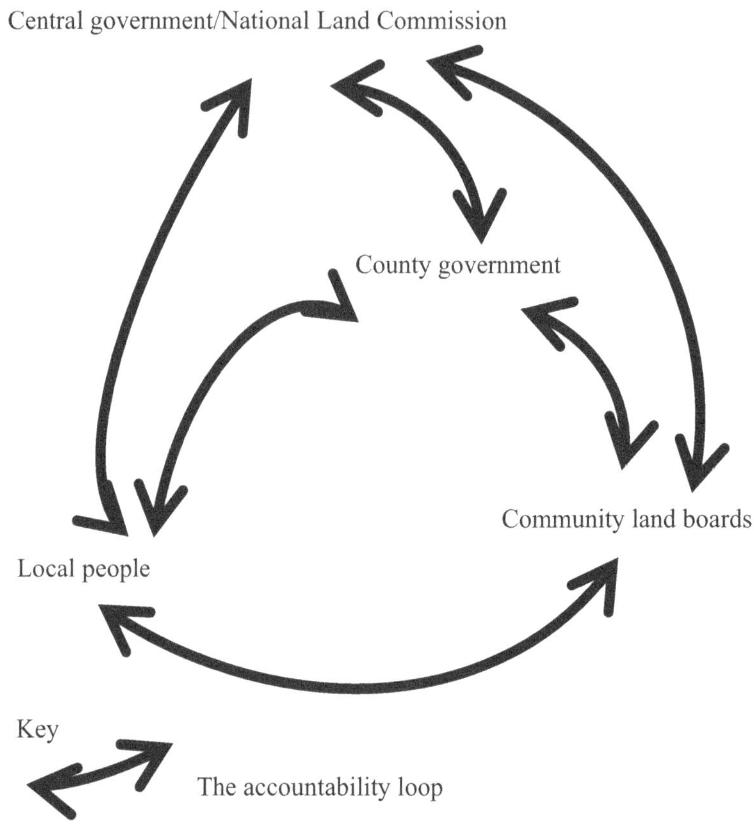

FIGURE 7.2 The Accountability Framework

Source: Mwangi, 2012

For the successful implementation of a decentralization framework, it has been argued that there is need to include political, institutional and fiscal dimensions, all which are key dimensions of decentralization. Analyses of other countries' case studies where land administration decentralization has been attempted reveal that rarely is true decentralization achieved; instead what has been the practice in these countries is a combination of deconcentration and devolution (Kauzya, 2007). The Narok County study revealed that as much as devolution is the preferred form of decentralization there is need for empowering decentralized functions at the local level through competent human capacity and appropriate funding. There is also need for creating democratic, locally accountable and specialized community land institutions so as to avoid devolving the very problems that land administration faces in a centralized system.

In conclusion, the author is of the view that for effective decentralization of land administration, it is imperative that the system put in place has clear and non-conflicting responsibilities and functions. Coupled with this both the national and local government have to commit to providing adequate budget to support the land administrative service delivery at the local levels.

To aid in this realization this author uses Kenya's institutional framework to propose a relationship model showcased in figure 7.1.

The details of the tasks to be included in the preceding relationships should be agreed on by all the relevant land sector stakeholders within a particular jurisdiction. It is proposed that this relationship model should be used together with an accountability framework shown in figure 7.2. As the framework indicates all the different actors are looped together and hence are accountable to the local people who are the consumers of the services. This demonstrates that for a decentralized land administration system downwards accountability is key. The people are also in their different capacities accountable to the institutional actors, especially in management of land resources at their disposal. The institutions, on the other hand, have to be accountable to each other with varying degrees of accountability. This makes programmes easy to sell to all stakeholders, thus making all want to be part of the decentralization process and eager to ensure the success of implementation.

References

Adams, M. (2000): *Breaking Ground: Development Aid for Land Reform*. London: ODI.

Agrawal, A. and Ribot, J. (1999): Accountability in Decentralization: A Framework with South Asian and West African Environmental Cases. *Journal of Developing Areas*, 1–63.

Ahmad, J., Shantayan, D., Khemani, S. and Shah, S. (2005): "Decentralization and Service Delivery" World Bank Policy Research Working Paper 3603, World Bank Development Research Group, Washington, DC.

Barraclough, S. (1999): Land Reform in Developing Countries, the Role of the State and Other Actors; United Nations research Institute of Social Development, Geneva.

Benjaminsen, T. A. and Lund, C. (2003): Formalisation and Informalisation of Land and Water Rights in Africa: An Introduction. In T. A. Benjaminsen and C. Lund (eds.), *Securing Land Rights in Africa*. London: Routledge.

Boore, C. et al. (2019): Land Law Reforms in Kenya: Devolution, Veto Players and the Limits of an Institutional Fix. In *African Affairs*; Oxford University Press; London.

Borras, Jr. S. M. and Franco, C. J. (2007): Land Policy and Governance: Gaps and Challenges in Policy Studies, OGC Policy Brief No. 1, Brasilia: UNDP International Poverty Centre.

Bruce, J. and Knox, D. (2009): Structures and Stratagems: Making Decentralization of Authority Over Land in Africa Cost-Effective. *World Development*, Vol. 37 (8), 1360–1369.

Burns, T. (2007): Land Administration Reform: Indicators of Success and Future Challenges World Agriculture and Rural Development Discussion Paper 37, World Bank, Development Research Group, Washington, DC.

Dale, P. and Mc Laughlin, J. (1988): *Land Information Management*. Clarendon Press, Oxford; Oxford University Press; New York.

Dale, P. and Mc Laughlin, J. (1999): *Land Administration*. Oxford University Press; Oxford.

Deininger, K. and Enemark, S. (2010): Land Governance and the Millennium Development Goals. In K. Deininger, C. Augustinus, S. Enemark and P. Munro-Faure (eds.), *Innovations in Land Rights Recognition, Administration, and Governance*. The World Bank; Washington, pp. xiii–xxvi.

Food and Agricultural Organization. (2007): Good Governance in Land Tenure and Administration; Rome.

Hillhorst, T. (2010): Decentralization, Land Tenure Reforms and Local Institutional Actors: Building Partnerships for Equitable and Sustainable Land Governance in Africa. *Land Tenure Journal*, 35–59.

Kauzya, J. (2007): "Political Decentralization in Africa: Experiences of Uganda, Rwanda, and South Africa": (Discussion Paper for the United Nations Department of Economic and Social Affairs), New York.

Kummer, D. (1992): *Deforestation in Post war Phillipines*. Chicago University Press; Chicago.

Land Development & Governance Institute. (2015): Status of Service Delivery in the Land Sector; 18th Scorecard Report 28th October 2015, Nairobi.

Marysse, S. (2004): "Decentralization Issues in Post-Conflict Democratic Republic of the Congo (DRC) (Preliminary Draft);" Institute of Development Policy and Management Discussion Paper 2004–2005, University of Antwerp, Antwerp.

Mwangi, W. (2012): Decentralization and Land Administration: Case Study of Narok District: A Research for Institution of Surveyors of Kenya, Nairobi.

Mwangi, W. and Nyika, D. (2018): *Land Administration, Principles and Processes*. Quinexx Publishers; Nairobi.

National Land Commission. (2016): Devolving Land Governance, 2015/2016 Annual Report; National Land Commission, Nairobi.

Onalo, P. L. (1986): *Land Law and Conveyancing in Kenya*. Heinemann Kenya; Nairobi.

Republic of Kenya. (2009): *Sessional Paper No.3 of 2009 on National land Policy*. Government Printer; Nairobi.

Republic of Kenya. (2010): *Constitution*. Government Printer; Nairobi.

Smoke, P. (2003): Decentralization in Africa: Goals, Dimensions, Myths and Challenges. *Public Administration and Development*, Vol. 23, 7–16. John Wiley and Sons; New York.

Syagga, P. M. (2006): Land Ownership and Uses in Kenya: Policy Preoccupation from an Inequality Perspective. In SID (ed.), *Readings on Inequality in Kenya*. Society for International Development; Nairobi, Kenya.

Yahya, S. and Swazuri, M. (2007): Customary Leaseholds and Perpetual Tenancies on the Kenyan Coast. Findings in Built and Rural Environments (FiBRE) Series.

UNECE. (1996): United Nations Economic Commission for Europe, Land Administration; Guidelines with Special Reference to Countries in Transition, United Nations.

Upton, C. (2009): Custom and Contestation: Land Reform in Post-Socialist Mongolia. *World Development*, Vol. 37 (8), 1400–1410, Oxford.

Wanjala, S. (2000): 'Recurrent Themes in Kenya's Land Reform Discourse Since Independence. In S. Wanjala (ed.), *Essays on Land Law: The Reform Debate in Kenya*. Faculty of Law, University of Nairobi; Nairobi.

8

THE POLICY FRAMEWORKS SUPPORTING SUSTAINABLE DEVELOPMENT OF REAL ESTATE IN KENYA

Mwenda K. Makathimo

1. Introduction

As a signatory of the Rio declaration at the United Nations Conference on Environment and Development in 1992 and the Sustainable Development Agenda 2030, Kenya has made tremendous strides in terms of legal and policy changes to align with and implement the commitments made. The changes have had the effects of changing how real estate developments are conceived, designed, implemented and managed.

The most notable legal and policy change is the full embrace of the tenets of sustainable development by anchoring them in the country's Constitution in the year 2010. Besides the Constitution, key statutory developments have been made. The statutes and attendant delegated legislation have addressed socio-economic dimensions; conservation and management of environmental resources; public and stakeholder participation including minorities and vulnerable groups; and implementation mechanisms.

In addition to legislative developments, Kenya has developed several overarching and sectoral policies articulating intentions and development plans that have mainstreamed the Rio principles. These policies range from those addressing infrastructural development; land use and management; environmental management; public health; physical planning; water resources management; energy management among others.

This chapter reviews these policy frameworks in the context of sustainable real estate development and management.

2. Sustainable Development Agenda

The Rio Declaration on Environment and Development made by most Members of the United Nations in 1992 set the foundation for sustainable development by providing the principles upon which development will be based. Principle three of the Declaration requires the right to development to be fulfilled in a manner that ensures equity in meeting development and environment needs of the present and future generations. The fourth principle emphasizes that for sustainable development to be achieved, environmental protection has to

DOI: 10.1201/9780429279256-8

constitute an integral part of the development process. Principle 11 requires states to enact effective environmental legislation, standards and management objectives that reflect their environment and development context.

Besides the Sustainable Development Principles, United Nations Member States have set development goals periodically. They first set the Millennium Development Goals (MDGs) which were replaced by Sustainable Development Goals (SDGs) in the year 2015. The SDGs have been outlined under the 2030 Agenda for Sustainable Development. The Agenda seeks to integrate and balance economic, social and environmental dimensions in development. It is within this background that Kenya set out to review its policy frameworks and development plans. The next section discusses the key frameworks influencing sustainable development and management of real estate in Kenya.

3. Constitutional Provisions

Kenya undertook comprehensive constitutional reforms in the year 2010 culminating in the promulgation of a new Constitution. The Constitution embraces and anchors the concept and principles of sustainable development. Right from the preamble it proclaims respect for the environment as a heritage and commits people to sustain it for the benefit of future generations. It places sustainable development, participation of the people, equity, social justice, inclusiveness, equality and non-discrimination as some of the national values and principles of governance[1](GOK, 2010). These principles and values are binding on every state officer, public officer, private persons and entities. They form a guide for every legislation, policy formulation, planning process, design and implementation of projects. It is therefore a requirement that any person or authority making real estate acquisition, design, development, management or disposal decisions abides by these values and principles.

Apart from the provisions on national values and principles of governance, the Constitution has secured key tenets of sustainable development under the Bill of Rights. Key among these rights is the right to a clean and healthy environment which includes the protection of these rights for future generations[2] This provision has significant bearing on all players in the real estate sector. Every development has to meet these criteria if it is to receive approval. Even if the development receives approval and it fails this test in any manner any person has the right to take legal action and compel the state to ensure compliance and seek compensation from the parties in breach.

Besides the environmental rights, the Constitution provides for economic and social rights. Every person has a right to accessible and adequate housing, social security, reasonable standards of sanitation, and clean and safe water.[3] The state is required to work progressively to ensure that this right is secured for every citizen. This provision has compelled the government to seek to encourage investments in the housing and infrastructure sectors. Real estate developers are required at the minimum to observe these requirements with the ones in the residential sectors expected to address the need for adequate housing while designing their projects.

For players in the real estate industry, it is particularly important to note that the Bill of Rights covers the protection of right to property. One is allowed to own property of any description either individually or in association with others.[4] This protection covers all interests in property. The protection is given against all forms of dispossession including those arising from the state. Should the property be required for public purpose, then prompt and just compensation is mandatory (GOK, 2010). This provision is important because it offers the much needed security for investments in real estate. With security of property rights guaranteed, property developers and managers can adopt and implement long-term objectives including those of investing in environment friendly infrastructure that may take long periods to procure positive financial returns.

The Bill of Rights under the Constitution further provides for equality and freedom from discrimination.[5] This provision outlaws all forms of discrimination. Real estate developments must therefore take into account the needs and challenges faced by all persons. For instance, the designs must provide facilities for people living with disabilities and ensure their safety and security in case of emergencies within buildings. Property managers and landlords must not practice discrimination on the basis of gender, race, marital status, origin, religion, culture, belief, language, disability or any other bases in letting of houses or facilities.

The right to access information held by the state and any other person that is necessary for the protection of fundamental rights and freedoms is also guaranteed by the Bill of Rights.[6] This is important for instance if a tenant or building user has to protect any of their fundamental rights. Property owners or managers may be compelled to provide information such as building plans, catalogues of materials used in construction or any other proprietary information. Similarly, property owners, developers and managers may invoke this provision to demand for information to protect their property rights or any other rights enjoyed under the Constitution.

Every person has the right to fair administrative action that is expeditious, efficient, lawful, reasonable and procedurally fair.[7] This by implication means that real estate developers are entitled to receive efficient services for instance when seeking building approvals or any other planning consents from government authorities. Any denial of such approvals and consents must satisfy the requirement of fairness. Property buyers are similarly entitled to receiving efficient property transfer and registration services from the relevant government authorities.

The Bill of Rights requires the state to provide access to justice for every person while ensuring that no prohibitive fee is charged if any.[8] This means that every person has a right to access court of law, be represented and be heard fairly. This is a critical provision especially in ensuring the enforcement of all the provisions relating to sustainable development. Real estate investors, developers, managers and users require a reliable and fair justice system to resolve any disputes during the development and lifecycles of buildings. Given that striking a balance between economic, environmental and social considerations is not always a matter of precision, robust, fair and efficient justice delivery institutions play critical roles in maintaining societal harmony.

Chapter five of the Constitution solely addresses Land and Environment. It enumerates principles of land policy[9] that include:

(i) equitable access to land
(ii) security of land rights
(iii) sustainable and productive management of land resources
(iv) transparent and cost effective administration of land
(v) sound conservation and protection of ecologically sensitive areas
(vi) elimination of gender discrimination in law, customs and practices related to land and property in land
(vii) encouragement of communities to settle land disputes through recognized local community initiatives consistent with the Constitution (GOK, 2010).

These principles have a strong correlation with principles of sustainable development encapsulated under the Rio Declaration on Environment and Development. Legislation, regulations, sectoral policies, standards and administrative practices on land are bound to follow these constitutional principles. All forms of real estate falling within the scope of definition of land in the Constitution must therefore be developed and managed in accordance with these principles.

In addition the chapter has enumerated the following as the obligations of the state in respect to environment[10]:

(i) ensure sustainable exploitation, utilization, management and conservation of the environment and natural resources, and ensure the equitable sharing of the accruing benefits
(ii) work to achieve and maintain a tree cover of at least ten per cent of the land area of Kenya
(iii) protect and enhance intellectual property in, and indigenous knowledge of, biodiversity and the genetic resources of the communities
(iv) encourage public participation in the management, protection and conservation of the environment
(v) protect genetic resources and biological diversity
(vi) establish systems of environmental impact assessment, environmental audit and monitoring of the environment
(vii) eliminate processes and activities that are likely to endanger the environment
(viii) utilize the environment and natural resources for the benefit of the people of Kenya (GOK, 2010).

All persons are required to cooperate with the state in discharging these obligations, conserving the environment and ensuring sustainable development. This provision takes sustainable development from just principles and values to mandatory obligations of all state organs, private organizations, communities and individuals. It forms the basis for moving the agenda from policy statements to actionable agenda that links the values and rights with obligations and responsibilities. The enforcement of these environmental rights and responsibilities[11] is also provided for through the court systems. It is worth observing that the Constitution has provided for the establishment of a special court for handling matters on Environment and Land with a status of the High Court.[12] To expound further on the provisions for values, principles, rights and obligations and to provide for their detailed implementation, the Constitution has directed Parliament to enact legislation on Land and the Environment.[13] The next section reviews the legislation relating to sustainable development of Real Estate in Kenya.

4. Legislation Affecting Sustainable Development of Real Estate

Kenya has enacted the framework environment law and other supportive legislation in line with principle 11 of the Rio Declaration (UN, 1992). The key legislation relating to sustainable real estate development include The Environmental Management and Co-ordination Act of 1999; The Physical and Land Use Planning Act, 2019b; The Public Health Act Cap 242, 2017a; Energy Act, 2019a; the Water Act, 2016c; The Land Act, 2012a; The Land

Registration Act, 2012b; the Community Land Act, 2016a; The Forest Management and Conservation Act, 2016b.

4.1 The Environmental Management and Co-ordination Act of 1999

The Environmental Management and Co-ordination Act (EMCA) was passed by parliament in 1999 and commenced on 14th January 2000. The Act was subsequently amended in the year 2015 to align it to the provisions of the Constitution of Kenya 2010 (The Constitution of Kenya (Amendment) Bill, Government of Kenya, 2015). This law had the effect of setting up institutions, standards and processes for environmental management and conservation.

It made it mandatory for Strategic Environmental Assessments (SEA) to be conducted on all policies, plans and programs.[14] This has the implication of requiring policies, plans and programs relating to real estate development to be assessed to ensure that they comply with sustainable development principles. This law makes it mandatory for all project proponents to apply for an Environmental Impact Assessment (EIA) License.[15] This by implication means all real estate development projects are subject to Environmental Impact Assessment (EIA) before they can receive approval. Before approval the EIA reports are subject to public participation. EMCA requires the summary of the report to be published in at least two newspapers circulating in the project area. The residents of the area and other stakeholders including state agencies are afforded an opportunity to submit their comments on the contents of the report and the anticipated impacts.[16]

Fresh EIA may be required to be submitted to the National Environment Management Authority (NEMA) by the project proponent if there is a substantial change or modification of the project. The fresh EIA may also be demanded at the cost of the proponent if inaccurate information was submitted at the point of seeking approval.[17] It must be noted however that NEMA can revoke the EIA license if the licensee contravenes the conditions given at the point of approval.[18] NEMA is expected to conduct environmental audits and monitor developments or activities on any land and premises that may have substantial impacts on the environment.[19] It is therefore required that real estate developers, owners and managers comply with the conditions of the EIA license during the project period and the life cycle of the development.

EMCA provides for the setting of environmental quality standards. The standards include water quality standards; effluent discharge standards; air quality standards; waste standards; standards on pesticides and toxic substances; radiation standards; and standards for control of noxious smells.[20] The detailed specifications of these standards have been published in the regulations made under the Act. All these standards apply during the real estate project implementation phase and also the development operation phase. Professional project teams have to heed to the standards while doing the designs and making the various construction material specifications. The contractors are also bound to adopt construction methods that comply with the standards. Property and facility managers are equally expected to abide by and ensure that building users and tenants adhere to the set standards.

EMCA further confers NEMA with the powers to issue environmental restoration orders, environmental easements and conservation orders.[21] It is important that property owners, developers and managers are well appraised with contents of such orders where they have been given to ensure compliance. In the case of environmental easements property owners are entitled to compensation where their rights and interest in property are adversely affected.

National Environment Tribunal (NET) is established under this statute to hear and make determination on matters arising from implementation of EMCA. Appeals against decision of NET lie with the Environment and Land Court.[22] Investors, owners, real estate professionals and the public are therefore provided with the mechanism for resolving disputes arising from the implementation of this statute. It must be borne in mind that failure to comply with the provisions of EMCA is regarded as an offence. Failure to exercise due diligence, efficiency and economy to ensure compliance by an agent, employee, servant or partner could result in conviction.[23] These convictions could lead to penalties and business costs besides attracting negative reputation.

4.2 The Physical and Land Use Planning Act, 2019

The Physical and Land Use Planning Act, 2019 is another important legislation that makes provisions directly relating to sustainable development of real estate in Kenya. It sets the objective of seeking to provide a framework for equitable and sustainable use, planning and management of land; ensuring that investments in property benefit local communities and their economies.[24] The statute sets values, principles and norms for guiding the planning processes.[25] The values, principles and norms are explicit that all physical and land use planning activities will seek to promote sustainable development; establish livable communities; integrate economic, social and environmental needs of the present and future generations; conserve land and natural resources; and preserve land with special functions, culture and heritage of the people (GOK, 2019b). These norms, values and principles affirm the values and governance principles in the Constitution of Kenya 2010.

Having set the objects, values, norms and principles, the statute proceeds to set physical and land use planning institutions at the national and devolved government levels.[26] The establishment of these institutions at the various levels, especially devolved levels, satisfies principle 10 of the Rio Declaration. The Act further outlines the various types of Physical and Land Use Plans, contents and preparation processes.[27] The Physical and Land Use Plans are categorized as:

(i) National Physical and Land Use Development Plans
(ii) Inter County Joint Physical and Land Use Development Plans
(iii) County Physical and Land Use Development Plans
(iv) Special Area Plans

It is important for real estate investors and practitioners to acquaint themselves with the various plans in relation to location of their projects since the plans will prescribe the type and scope of developments allowable and their relationship with the larger regions. This would enable them to conduct project feasibilities with efficacy and leverage opportunities created by planned complementary infrastructure and other developments.

In addition to the types of plans, the Act also outlines the development control objectives and processes.[28] Development control under this law aims at ensuring orderly development; optimal land use; proper execution of approved plans; protection and conservation of the environment; promoting public health and safety; safeguarding national security; and participation of the public in decision making (Ibid).

The Act gives County Governments powers to exercise development control for areas within their jurisdiction.[29] The development control powers include:

(i) prohibit or control the use and development of land and buildings in the interest of proper and orderly development
(ii) control or prohibit subdivision of land
(iii) consider and approve all development applications and grant all development permissions
(iv) ensure the proper execution and implementation of approved physical and land use development plans
(v) formulate bylaws to regulate zoning in respect of use and density of development
(vi) reserve and maintain all land planned for open spaces, parks, urban forests and green belts in accordance with the approved physical and land use development plans
(vii) consider and determine development planning applications made in respect of land adjoining or within vicinity of safeguarding areas (GOK, 2019b).

These development control powers are to be exercised subject to the County Government Act, 2012 and the Urban Areas and Cities Act 2011 (Ibid). These two Acts make complementary provisions with respect to county planning and allow for integration of planning functions as well as delegation to lower levels within a county.[30] It is critical for players in the real estate industry in Kenya to understand this important role of development control played by the County Government institutions. No project approval and successful implementation can be legally realized without due compliance with the regulations that they set.

There are dispute resolution mechanisms established under the Physical and Land Use Planning Act. At the national level there is established the National Physical and Land Use Liaison Committee. This committee has jurisdiction to hear and determine claims made against decisions of the National Planning Authority regarding development of major infrastructure; reserving of public land for projects; implementation of Inter County Physical and Land Use Plans; and environmental impact on ecologically sensitive areas by implementation of strategic project.[31]

At the devolved level the dispute resolution institution is the County Physical and Land Use Planning Liaison Committee. It has jurisdiction over complaints and claims against applications submitted to the county planning authority, appeals against decisions of county planning authority and enforcement notices by the County Government.[32] If the aggrieved parties are dissatisfied with the decisions of the liaison committees at the county or national levels, they have a right to appeal to the Environment and Land Court established under the provisions of the Environment and Land Court Act of 2011 and the Constitution.

4.3 The Public Health Act Cap 242

The Public Health Act Cap 242 is another key legislation in relation to sustainable development and management of real estate. Its provisions relate to the Sustainable Development Goal on good health and wellbeing (UN, 2015). The Act makes provisions that have implications of real estate development during the phases of design, planning, construction, occupation, management, maintenance and demolition.

It defines and prohibits nuisances while authorizing the County Governments (local authorities) to ensure cleanliness and prevention of nuisances in relation to housing and sanitation.[33] Nuisances are defined in a detailed manner in the Act.[34] It is however important to observe that the scope covers, dirt, polluted water supplies, noxious or waste matter, flow of effluent, accumulation of refuse, accumulation of materials including building materials that can harbor vermin, offensive smells; buildings that are poorly constructed or maintained in an unsafe or injurious manner, without proper water supply, ventilation; any area of land in a state that can cause infectious, communicable or preventable disease (GOK, 2017a). In sum any act or omission which may be dangerous to life or injurious to health is prohibited on any land, premises or building.

The Act empowers medical health officers to inspect, examine, give notices to property owners, take measures to remedy the cause of any nuisance including demolition of any dangerous dwelling units.[35] The statute further gives powers to the cabinet secretary to make rules to guide the enforcement of provisions relating to housing and sanitation. The scope of the rules would cover inspection of land and buildings; construction of buildings including ventilation and lighting; periodical painting and white washing; drainage, discharge of waste and affluent; subdivision of land and layout of sites; use and number dwellings and occupants; and keeping of animals among others.[36]

County Governments (local authorities) are required under the Public Health Act Cap 242, to make bylaws with respect to buildings and sanitation. The bylaws would cover issues such as controlling of construction of buildings and materials to be used; height of buildings and chimneys; building space, ventilation, dimensions of rooms intended for human habitation; provision of fire escapes; issuance of occupancy certificates; repair or demolition of dilapidated buildings; water supply systems and standards fittings drainage and waste disposal systems; regulating sewers and private drains; prohibiting projections and obstructions in front of buildings and streets.[37] The County Governments are further bestowed with powers to pass or reject building plans, and require removal or alteration of building works in case of breach of bylaws during or after construction.[38]

It is significant to observe that the medical officer of health or any health authority permitted by the Act has power to carry out inspections; cleanse and disinfect a building or a premise; and stop the occupation of a building premises or a dwelling place for purposes of prevention or suppression of an infectious disease.[39] It should be noted that the Act prohibits building owners and property managers from letting infected houses or failing to disclose information regarding infection of houses. Failure to comply would attract penalties and conviction.[40]

4.4 The Energy Act of 2019

Kenya has enacted the Energy Act of 2019 that provides a framework for development, supply and management of affordable and clean energy in line with Sustainable Development Goal number seven (UN, 2015). The Act makes it a government obligation to facilitate the provision of affordable energy. It requires the cabinet secretary in consultation with stakeholders to develop a national energy policy; prepare an integrated national energy plan with inputs of energy service providers and County Governments; and monitor the implementation of the plan.[41] The cabinet secretary is further charged with the duty of promoting the development, distribution and use of renewable energy technologies such as wind, solar, tidal waves, hydropower, biogas and municipal waste.[42]

The Act establishes various institutions involved in energy development and management including the Energy and Petroleum Regulatory Authority (EPRA); Rural Electrification and Renewable Energy Corporation (REREC); and Energy and Petroleum Tribunal (EPT).[43] These institutions play various roles and have powers that have significance in the development and management of real estate.

EPRA is given the function of preparing and ensuring the implementation of National Energy Efficiency and Conservation Plan, and recommending to the cabinet secretary regulations on energy efficiency and conservation. It is bestowed with power to designate factories and buildings for the purpose of energy conservation; collect information from building owners on energy conservation; enforce regulations; regulate the conduct of energy audits; accredit energy auditors; specify energy conservation standards and benchmarks; authorize electrical installations; and certify electrical contractors and workers among other functions.[44]

The REREC has the functions of overseeing rural electrification program; developing and updating rural electrification master plans; developing and updating renewable energy master plans; and developing, promoting and managing the use of renewable energy technologies among other functions under the Act.[45] NET on the other hand has jurisdiction to hear and determine all matters brought to it regarding energy and petroleum save for criminal matters.[46]

Besides the institutions established under the Energy Act, County Governments are given powers to ensure efficient use of energy and its conservation. They may amend energy conservation building codes to suit local climatic conditions; direct building owners and occupiers to comply with energy efficiency and conservation codes; and create awareness on efficient use and conservation of energy. County Governments are also given powers to inspect buildings, issue directions and enforce standards on energy efficiency and conservation. In addition, they have powers to make rules to facilitate the effective performance of these functions provided that they are consistent with the provisions of the Energy Act 2019 and the Constitution of Kenya 2010.[47]

Building developers, owners, occupiers and managers are required to comply with the provisions of this Act. They particularly have the duty to ensure electrical installation works are carried out by qualified and certified workers and contractors and that they meet the standards set and approved under the Act.[48] Owners, occupiers and property managers are also required to conserve energy, facilitate energy audits and conduct consumption analysis. They are also expected to provide information and records required by EPRA or County Governments for purposes of efficient use and conservation of energy. Failure to comply with these requirements would attract penalties and fines.[49]

The Act makes provisions with respect to acquisition of rights of way, wayleaves and use of land for development of energy resources and infrastructure. The acquisition of these rights and wayleaves may result in restrictions, prohibitions of owners' rights or compulsory acquisition of interests in land. The Act requires compensation to be made to the affected owners affected by the restrictions, prohibitions or compulsory acquisition[50]

4.5 The Water Act 2016

The other statute that has relevant provisions that have impact on sustainable development and management of real estate is the Water Act 2016. It was enacted with the purpose of regulation, management and development of water resources, water and sewerage services.[51]

It empowers the cabinet secretary to formulate a National Water Strategy every five years. The strategy should detail plans and programs for protection, conservation, control and management of water resources.[52] This strategy is to be implemented by the Water Resources Authority (WRA) which is established under the Act[53] and all the stakeholders. WRA has the duty to formulate and enforce standards and regulations on management and use of water resources; give permits for use of water from a water resource; construction of waterworks; creation of easements; abstraction of ground water among others. It has powers to prohibit activities on catchment areas, and make or permit entry into land for purposes of inspection or carrying out of works.[54] It is therefore necessary that real estate developers, owners and managers appraise themselves with the water resources management strategies, the standards and regulations, conditions and procedures for obtaining permits; and the prohibitions put in place for purposes of water resources protection under the provisions of this Act.

The subject of water services provision is premised on the right to clean and safe water in adequate quantities and reasonable standards of sanitation as provided for by the Constitution of Kenya 2010, Sustainable Development Goal number six (UN, 2015) and reinforced by the Act.[55] In line with this provision the cabinet secretary is required to prepare a National Water Services Strategy every five years setting the standards and plans for realization of water rights.[56] The Act has provided for establishment of Water Works Agencies to undertake development, maintenance and management of national water works; operate the water works; and provide water services, technical and capacity building services to counties.[57]

The statute also establishes the Water Services Regulatory Board (WSRB) with the purpose of protecting the interests and rights of consumers in the provision of water services. WSRB has the duties of prescribing the national standards for provision of water services and development of assets for water service providers; registration and licensing water service providers; evaluating and recommending water tariffs to county water service providers; monitoring compliance with standards of design, construction, operation and maintenance of facilities for the provision of water services.[58] It is also key to note that the Act allows County Governments to establish water service providers with the approval of WSRB to provide services within a specific area and develop county assets for water provision.[59] Disputes arising out of implementation of the various mandates under the Water Act are handled by the water tribunal. Any appeals against the decision of the tribunal fall under the jurisdiction of the Environment and Land Court.[60]

The provision of water services is critical to sustainable development and operation of real estate developments. It is essential that all practitioners and stakeholders in the real estate industry fully appreciate the standards and regulations set by the various institutions; their rights with respect to service provision and fair tariffs and take their due responsibilities as required by the Act.

4.6 The Forest Management and Conservation Act 2016

The Forest Management and Conservation Act 2016 has also made a provision supporting sustainable real estate development. It requires every county to make it a condition before approval, for every housing estate developer to make provision for establishment of green zones at the rate of at least 5 percent of total land area of the intended development. The Act also commits the Kenya Forest Service to give technical capacity to the county and the developers to establish green belts.[61]

Kenya has enacted several land laws that set principles, frameworks and processes for sustainable land management.

4.7 The Land Act 2012

The Land Act 2012 provides the guiding values and principles for land management that include security of land rights, non-discrimination, efficiency, inclusivity, accountability, equity and sustainability among others.[62] It outlines forms of tenure to include freehold, leasehold, customary and other forms defined by statute. Additionally, the Act provides the methods of acquisition and dispossession of title to land.[63] The statute provides for the processes of management of public land; administration of leases, licenses and agreements on public land; contracts transfers and transmissions; settlement programs; creation of rights of way and compulsory acquisition of rights to land (GOK, 2012a).

4.8 The Land Registration Act 2012

The Land Registration Act 2012 establishes a framework and processes for registration of rights and transactions on land. It details how registers and registries are to be created and maintained. The Act also covers the subjects of mapping and determination of parcel boundaries; issuance of certificate of title, leases and search documents. It also addresses dispossessions affecting land; charges, transmissions and trusts; and creation of easements and analogous rights (GOK, 2012b).

4.9 The Community Land Act of 2016

Another important legislation on land is the Community Land Act of 2016. This Act provides for the recognition, protection and registration of community land rights; management frameworks for community land; the roles of the County Government in relation to unregistered community land; Nature of Community title; and conversion of community land to either public or private land.

The legislation on land provides a base upon which the bundles of real estate rights are established, secured and managed. This provides certainty and hence confidence required to spur investments on land for current and future generations.

5. Sectoral Policies

Besides the constitution and legislation discussed earlier, there are other sectoral policies in form of sessional papers that influence development of real estate in Kenya. These include the Sessional Paper number three of 2009 on National Land Policy; Sessional Paper number one of 2017 on National Land Use Policy; The National Environment Policy of 2013.

The National Land Policy seeks to guide the country towards efficient, sustainable and equitable access, management and use of land (GOK, 2009). It's worth observing that this policy was passed before the promulgation of the Constitution of Kenya 2010. The legal and institutional reforms it seeks have mainly been addressed by the Constitution and legislation discussed earlier. The National Land Use Policy has the objective of providing a guide for environmentally and socially responsible land use in Kenya. It also seeks to promote practices that ensure optimal land utilization in a sustainable manner (GOK, 2017b). It aligns with the provisions of the Constitution of Kenya 2010. The National Environment Policy provides a framework for an integrated approach to planning and sustainable management of the environment. It aims at enabling the country to achieve a better quality of life for the present and future generations (GOK, 2013). It equally aligns with the constitutional provisions on matters to do with the environment.

Conclusions

Kenya has put in place legal and policy frameworks for supporting sustainable real estate development and management. The constitutional provisions include values and principles on governance; the Bill of Rights that includes property rights, environmental and social economic rights; and legislative, enforcement and dispute resolution institutions. The legislative frameworks are aligned with the constitutional principles, values and foundations. They cover environment, planning, public health, energy, water and land management. The legislation gives detailed operational processes together with the implementation institutions. In addition to the legislation, sectoral policies on environment, land governance and land use outlining government objectives and operational commitments have been formulated and adopted by parliament.

These legal and policy frameworks have significant implications on how real estate is acquired, developed, used and managed. They have established basis for security of property, social and environmental rights; prescribed norms and values for sustainable development that correlate with Rio principles and the Sustainable Development Goals. They set standards, regulations and administrative codes that form a sound basis for approving, sanctioning or prohibiting action at the various points of real estate development and management cycle with the aim of securing benefits for current and future generations in an equitable manner. The challenge that remains is the effective and efficient implementation of these frameworks and making the due practice and cultural changes while remaining adaptable to the ever changing operational environment.

Notes

1 Article 10, Constitution of Kenya, 2010
2 Article 42, Ibid.
3 Article 43, Ibid.
4 Article 40. Ibid.
5 Article 27, Ibid.
6 Article 35, Ibid.
7 Article 47, Ibid.
8 Article 48, Ibid.
9 Article 60, Ibid.
10 Article 69, Ibid.
11 Article 70, Ibid.
12 Article 162, Ibid.
13 Articles 68 and 72, Ibid.
14 Section 57 A, Environmental Management and Coordination Act, 2000
15 Section 58, Ibid.
16 Section 59,60, Ibid.
17 Section 64, Ibid.
18 Section 67, Ibid.

19 Section 68, 69 Ibid.
20 Part viii, Ibid.
21 Part1x, Ibid.
22 Part xii, Ibid.
23 Part xiii, Ibid
24 Section 3, Physical and Land Use Planning Act, 2019.
25 Section 4, 5, Ibid.
26 Part ii, Ibid.
27 Part iii, Ibid.
28 Part iv, Ibid.
29 Section 56, Ibid.
30 Part xi, County Government Act, 2012
31 Sections 73,74, 75, Physical and Land Use Planning Act, 2019.
32 Section 76, 77, 78, Ibid.
33 Part ix, Public Health Act, Cap 242.
34 Section 118, Ibid.
35 Section, 117, 119, 123, 124 Ibid.
36 Section 126, Ibid.
37 Section 126 A, Ibid.
38 Section, 126C, 126 D, Ibid.
39 Section, 21, 22, Ibid.
40 Section, 30, Ibid.
41 Section 4, 5, 6, 7, Energy Act, 2019.
42 Section 75, Ibid.
43 Part iii, Ibid.
44 Section 10,148,149,150,187, 188, 200, 201, 203, 204, Ibid.
45 Section 44, Ibid.
46 Section36, Ibid.
47 Section, 193, 194, 195, 196, 199, Ibid.
48 Section, 151, Ibid.
49 Section 189, Ibid.
50 Section 170, 171, 172, 175, 179, 183, Ibid.
51 Section 3, Water Act, 2016.
52 Section 10, Ibid.
53 Section 11, Ibid.
54 Section 12,23,36,39,56,57,58,59 Ibid.
55 Section 63, Ibid.
56 Section 64, Ibid.
57 Section 65, 68, Ibid.
58 Section 70, 72, Ibid.
59 Section 77, Ibid.
60 Section 119, 124, Ibid.
61 Section 37, Forest Conservation and Management Act, 2016
62 Section 4, Land Act, 2012
63 Section 5, 7, Ibid.

References

Government of Kenya, (2000), Environmental Management and Coordination Act, National Council for Law Reporting, Nairobi, Kenya.

Government of Kenya, (2009), Sessional Paper No.3 on National Land Policy, Government Printer, Nairobi, Kenya.

Government of Kenya, (2010), The Constitution of Kenya, Government Printer, Nairobi, Kenya.

Government of Kenya, (2012a), Land Act, National Council for Law Reporting, Nairobi, Kenya.

Government of Kenya, (2012b), Land Registration Act, National Council for Law Reporting, Nairobi, Kenya.

Government of Kenya, (2013), National Environment Policy, Ministry of Environment and Natural Resources, Nairobi, Kenya.

Government of Kenya, (2016a), Community Land Act, National Council for Law Reporting, Nairobi, Kenya.

Government of Kenya, (2016b), Forest Management and Conservation Act, National Council for Law Reporting, Nairobi, Kenya.

Government of Kenya, (2016c), Water Act, National Council for Law Reporting, Nairobi, Kenya.

Government of Kenya, (2017a), Public Health Act Cap 242, National Council for Law Reporting, Nairobi, Kenya.

Government of Kenya, (2017b), Sessional Paper No.1 National Land Use Policy, Government Printer, Nairobi, Kenya.

Government of Kenya, (2019a), Energy Act, National Council for Law Reporting, Nairobi, Kenya.

Government of Kenya, (2019b), The Physical and Land Use Planning Act, Government Printer, Nairobi, Kenya.

United Nations, (1992), Report on the United Nations Conference on Environment and Development, United Nations, New York, USA.

United Nations, (2015), Transforming Our World; The 2030 Agenda for Sustainable Development, United Nations, New York, USA.

9

AN ANALYSIS OF THE EVOLUTION OF THE SUB-SAHARAN AFRICAN RESIDENTIAL REAL ESTATE RESEARCH

Aly Karam, François Viruly, and Taiwo Afinowi

Introduction

The academic interest and body of knowledge on sub-Saharan African housing markets is growing and becoming more diversified in tackling the real estate market topics. Invariably, treating the entire continent as a single research agenda is problematic, as sub-Saharan countries vary in their level of economic development, housing policies and the maturity of legal as well as economic and institutional arrangements. Because the legal and market structures in Anglophone and Francophone countries differ considerably, this review is only concerned with Anglophone sub-Saharan Africa.

The review covers the literature on residential property markets starting in the early 1970s when the first article on property in a sub-Saharan African city was published in the journal *Town Planning Review*, and therefore does not consider equally important literature that has developed around the commercial property sector during the same period. The literature under review has been disseminated through several highly rated real estate academic journals. This research is based on a comprehensive analysis comprising over one hundred academic journals spanning the period 1971–mid 2021. These articles are categorized chronologically and thematically to identify the development of the body of knowledge and the evolving research agenda in Anglophone Sub-Saharan countries. It is also an excellent opportunity to identify the quantity of literature available on housing properties discussing issues of importance to the sub-Saharan African countries.

Determining the Research Agenda in Sub-Saharan African Real Estate

This review attempts to identify the chronological evolution of the research agenda in sub-Saharan Africa. Over the past 20 years, there have been several articles evaluating different aspects of real estate research, and these articles mainly concentrated on the research published on the real estate markets of the United States. We decided to look at these studies to assist us in determining some of the important journals in the field that should be included in our

DOI: 10.1201/9780429279256-9

search for literature on sub-Saharan Africa. Some other work on studying trends in the real estate literature included Clauretie and Daneshvary (1993), who ranked the authors published in real estate journals by their number of publications. The research was based on three journals: *The Journal of the American Real Estate and Urban Economics Association*, *The Journal of Real Estate Research*, and *The Journal of Real Estate Finance and Economics*. It concludes that there is a significant difference between the group of academics who publish in real estate journals only and those who publish in real estate-finance journals. Their findings suggest that many of the academics who regularly publish in real estate journals are often not real estate academics but finance academics who publish on real estate related issues in real estate journals. Although, the notion of finance academics, taking the lead in real estate publications, has changed considerable as shown in this review.

In other types of research on literature, Diaz *et al.* (1996) attempted to rank real estate journals by their prominence in promotions and tenures. The research considered 30 journals that were correlated with the promotions and tenure for professors in the real estate field of research. The survey concludes that *Real Estate Economics (REE)* is the leading journal for academic output in the real estate sector. Moreover, *The Appraisal Journal (TAJ)* was the leading industry outlet for publications. So, the fact that an academic can publish in these two journals, the higher the probability of them getting a promotion or tenure. So, these journals were also included in our list of journals to search for publications on African housing real estate.

Webb and Albert (1995) concluded that *Real Estate Economics (REE)* was perceived by finance academics as the highest-ranked journal in the field of real estate, based on their perceived quality compared with other journals in the field. This was followed by the *Journal of Real Estate Research (JRER)*, *The Journal of Real Estate Finance and Economics (JREFE)*, *The Journal of Urban Economics (JUE)* and *Land Economics (LE)*. *The Appraisal Journal* (TAJ) was ranked 7$^{th.}$ and the Journal of Real Estate Literature was ran^{ke}d 12th. Redman *et al.* (1998) used citations to rank journals. Their ranking largely supported the work of Webb and Albert (1995), giving the highest-ranking to *Real Estate Economics (REE)* being the most cited journal. *The Journal of Real Estate Finance and Economics (JREFE)* came in second, followed by *The Journal of Real Estate Research (JRER)*. These findings were confirmed by Gibler and Ziobrowski (2002), which found the same order of journals where most of the literature originated, *REE*, *JREFE* followed by *JRER*. Later, this was also confirmed by Hardin *et al.* (2006).

Simons and Karam (2012) undertook the only work that has attempted to survey the journals read by academics, practitioners, and other *African Real Estate Society* annual conference attendees. Table 9.1 suggests that the top three journals considered by attendees for publication in Africa correspond to the conclusions reached by Weeks *et al.*, 2007; Redman *et al.*, 1999; Webb and Albert, 1995. Apart from the journals captured in Table 9.1, a few other articles are considered within scholarly works on the residential real estate market in sub-Saharan African; all articles are fully captured in the list of references.

Method

The analysis is based on an assessment of academic articles and excludes thesis, dissertations, monographs, and chapters in books. The initial search terms were real estate, housing finance, property finance, real estate management, valuation, and land tenure. They were all suffixed or prefixed with either Africa or a sub-Saharan African country. A total of 201 academic articles were found relevant, and the relevant articles were sourced. Then articles related to

TABLE 9.1 Comparative between top journals based on four different pieces of research.

Ranking	*This Research at AfRES 2008*	*Weeks, Finch and Hardin 2007*	*Redman Manakyan and Tanner 1999*	*Webb and Albert 1995*
1	*Journal of Real Estate Research*	*Real Estate Economics*	*Real Estate Economics*	*Real Estate Economics*
2	*Real Estate Economics*	*Journal of Real Estate Research*	*Journal of Urban Economics*	*Journal of Real Estate Research*
3	*Journal of Real Estate Finance and Economics*	*Journal of Real Estate Finance and Economics*	*Journal of Real Estate Finance and Economics*	*Journal of Real Estate Finance and Economics*
4	*The Appraisal Journal*	*Journal of Real Estate Literature*	*Journal of Real Estate Research*	*Journal of Urban Economics*
5	*Journal of Real Estate Portfolio Management*	*Journal of Housing Economics*		*Journal of Land Economics*
6	*International Real Estate Review*	*Journal of Urban Economics*		*Journal of Housing Research*
7	*Journal of Real Estate Practice and Education*	*Journal of Real Estate Portfolio Management*		*The Appraisal Journal*
8	*Journal of Real Estate Literature*	*Journal of Real Estate Practice and Education*		*REAA*
9	*Journal of Property Management*	*Journal of Land Economics*		*REI*
10	*Journal of African Real Estate Research*	*Urban Studies*		*Journal of Real Estate Research*
11	*Urban Studies*	*Journal of Regional Science and Urban Economics*		*Real Estate Finance Journal*
12	*Housing Studies*	*Journal of Property Research*		*Journal of Real Estate Literature*
13	*Development Southern Africa*	*Journal of Property Investment and Finance*		
14	*ActaStructilia*	*Journal of Housing Research*		

Source: Authors' construction

housing real estate were isolated, and the rest excluded, which yielded 72 articles for reviewing process.

The analysis was undertaken based on two broad approaches. The first considered the geographic extent of the research, paying particular attention to the countries that have dominated the sub-Saharan African research agenda; the second approach turns to the real estate topics tackled by researchers. The focus of the research did not necessarily identify whether African academics had undertaken the research but instead focused on whether the subject matter was focused on the sub-Saharan African continent.

The Geographic Extent of the Research

From a chronological perspective, the first identified article was published in 1971 by Onibokun, G.A; four articles appeared in the 1980s: in 1981, then 1986, 1988, and 1989. This number illustrated the relatively low level of research activity that occurred in sub-Saharan Africa. In total over the past 50 years, Nigeria saw the highest number of articles with 51 articles, the distant second being Ghana with 27 and South Africa comes in third with 26 articles. As illustrated in Table 9.2, a number of other countries also contributed one or two articles. As for the articles on residential real estate, there are even fewer; there were 72 articles in total for sub-Saharan Africa, and these are the ones that we will be discussing in this paper.

Analysis of Themes Chronologically

The analysis also connected the themes that were handled in the identified journal articles. The residential/housing sector dominated the published articles with 72 articles out of 201 articles on sub-Saharan African real estate. These arcticles address topical issues on housing market, education, facility management, research agenda for real estate, and ethics. In this chapter, we focus on articles on residential real estate.

The first article considered in the analysis was published in 1971 by Onibokun (1971) and focused on the rapid level of urbanization, lack of housing, and high densities found in African countries. Onibokun (1971) concludes that there is little finance available from the federal, state, or local government to support housing production. Also, the Western Nigeria Housing Corporation was unable to reach the poor in terms of housing finance, and this lack of finance will lead to urban growth with poor housing. This challenge of housing finance, in turn, leaves the poor in bad living conditions.

During the late 1980s, the focus was largely on the Nigerian property market. Ubogu (1988) considers the financial measures that have been implemented in Nigeria and their effect on the demand for loans. A further consistent theme in the literature is the poor development of financial institutions in sub-Saharan African countries and generally poor access to loans for housing. Megbolugbe (1989) studied the housing markets in Jos, Nigeria, using a hedonic index model. The author attempted to improve on previous research models by incorporating specific data on market conditions from an African case study perspective. While much of the focus was on financing, little attention seems to have been paid to the housing of the poor. Two articles are on finance, and the third article considered the valuation of residential properties.

The geographic breadth of the literature widens in the 1990s, with the focus shifting beyond Nigeria and onto other sub-Saharan African countries such as Zimbabwe, Kenya, and South Africa. The subjects handled by these papers also become more varied. Six articles considered the finance of housing, two considered housing valuation, and one was on housing markets. The articles dealing with housing finance (Rakodi, 1995; Smets, 1996; Kamete, 1997; Okonkwo, 1998; Tomlinson, 1999) consider issues that recent academic articles are still concerned with, namely housing finance for the lower-income groups. Therefore, it can be concluded that in the 1980s and 1990s concerns were on low-income housing finance, poor housing conditions, and affordability on the sub-Saharan African continent (Arimah, 1992; Arimah and Adinnu, 1995; Akpom, 1996; Ukoha and Beamish, 1997). Although, these

TABLE 9.2 Number of publications per year showing the countries each year.

Year	*Number*	*Residential*	*Country in Article*
1971	1	1	Nigeria
1988	2	1	Nigeria
1989	1	1	Nigeria
1992	1	1	Nigeria
1994	1		Ghana
1995	2	2	Nigeria and Zimbabwe
1996	2	2	Nigeria and South Africa
1997	2	2	Nigeria and Zimbabwe
1998	2	1	Nigeria and Kenya
1999	1	1	South Africa
2000	3		Kenya, 2 South Africa
2001	3	3	Nigeria and 2 South Africa
2002	3	1	2 South Africa, Sub-Saharan Africa
2003	6	1	Ghana, 2 Namibia, 3 South Africa and Kenya
2004	3	1	Uganda, Nigeria, Ghana
2005	5	3	Ghana, Namibia, Sub-Saharan Africa, 2 South Africa
2006	4	1	Nigeria, Africa, Ghana, South Africa
2007	19	6	9 Nigeria, 4 Ghana, Africa, 3 South Africa, 2 Malawi
2008	10	4	5 Nigeria, 2 Ghana, Kenya, 2 South Africa
2009	9	1	3 Nigeria, 3 Ghana (theoretical), South Africa, Africa
2010	11	4	4 Nigeria, 2 Ghana, Namibia, 3 South Africa, Africa
2011	11	3	6 Nigeria, 2 (theoretical), 2 South Africa, Africa
2012	8	4	4 Nigeria, 2 Ghana, Namibia, South Africa
2013	5	1	Nigeria, Namibia, Rwanda, 2 Ethiopia
2014	1	3	Ethiopia
2015	6	6	Namibia, Uganda, 3 Nigeria, Ghana
2016	3	3	3 Nigeria
2017	3	3	2 Ghana, Kenya
2018	6	4	Namibia, 2 Nigeria, 2 Ghana, Gambia
2019	4	4	3 Ghana, Namibia
2020	3	3	Nigeria, 2 Ghana
2021	3	1	Malawi, Tanzania, Uganda
	144	72	

Source: Authors' construction

concerns are still valied in Africa and other continents as reflected in the global housing agendas such as the 1998 United Nations Global Strategy for Shelter towards the millenlium, the Sustainable Development Goal 11, and the New Urban Agenda (UN-Habitat, 1996, 2015, 2016).

In 2000, there were no articles related to housing and real estate. 2001 saw three articles on this topic being published. Two of three were about low-income housing finance (Merrill, 2001; Moss, 2001). The third article not only talks about low-income/affordable housing finance but discusses the development of an "enabling housing environment" inclusive of finance, building material, regulations, and the role of different sectors in supplying the market with low-income housing (Ogu and Ogbuozobe, 2001). Again, the articles published suggest that housing real estate research revolves around housing for the poor.

In 2002, there was one article published on housing finance. It dealt with the errors and efforts of the South African government in finding different financial models for low-income borrowers (Tomlinson, 2002). In 2003, there was a similarity in the focus of articles published. There was one article on housing that showed that the South African government has not been successful in getting funds and mortgages to the low-income populations, and this failure has increased the frustrations of the aforementioned population (Karley, 2003).

Uganda produced the first article on housing in 2004. The article uses data on house prices and rentals taken from the 1999/2000 Uganda National Household Survey to test the efficacy of the housing market in all of Uganda. The authors, Knight et al. (2004), showed that there was a vibrant market in Uganda and that future expectations regarding the market were promising.

Gated communities as one of the neighborhood housing typologies have received considerable attention since the late 1970s in different parts of the world (Grant, 2005). Gated communities started appearing on the African continent in the 1980s. They were predominantly in South Africa and then started spreading over the rest of the African continent. In 2005, the first article analyzing their effect on the property markets in Ghana was published. The author, Richard Grant (2005), showed that the buyers of the houses in gated communities all lived abroad, or were still living abroad; hence globalization had arrived to the property market in Ghana. This phenomenon has shown and widened the gap in the Ghanaian society between people with money and people who do not have money. This paper is the beginning of research on how the global phenomena of gated communities took hold of local property markets in Africa.

South Africa had its first free democratic elections in 1994, which came with high expectations in the possible reform of the property market. In 2005, research was conducted on a small-town property market in Margate, in KwaZulu-Natal province testing whether there was any evidence of racial integration in residential areas (Lemon and Clifford, 2005). The research aimed to determine whether black populations were moving into previously white-only neighborhoods in small towns. The research concluded that only when economic integration occurred would residential property markets start to integrate.

The third paper on housing in real estate was published in 2005 and dealt with second home tourism (what can be termed leisure homes or holiday homes) in South Africa. They are used during weekends and holidays and are also held as an investment and sometimes as future retirement places. The research found that the South African second homes tourism market follows the same model as their international counterparts; they are for holiday migration rather than weekend usage (Hoogendoorn et al., 2005). They are also owned by a certain population living in larger cities, and these homes are in exclusionary areas; they do not encompass the poor.

The banking system in South Africa is generally considered by Kajimo-Shakantu and Evans (2006) to be one capable financially to assist in loans for low-income housing. Surprisingly, 12 years after apartheid, questions around banks' role in financing low-income housing were still being asked (Kajimo-Shakantu and Evans, 2006). In many African countries, banks could not play a role in financing low-income housing due to the lack of savings and a solid financial system. However, South Africa had sufficient finances and savings and had a financial system that could adapt to low-income housing.

The year 2007 saw an explosion of articles on sub-Saharan African real estate issues, with 19 articles on publication. Of all these articles, there were six articles published on housing. Quayson (2007) looked at the primary mortgage market initiative that was implemented in 2006 in Ghana. This helped in increasing the number of residential lenders to three banks, with other banks looking to join the initiative. The initiative showed the importance of the government's commitment to sustaining a housing finance market. It also brought confidence and security to the lenders knowing that the government is willing to put in the regulatory framework to support them.

Asabere (2007) looked at the problems that rent control units caused in the value of the housing within the township of Tema in Ghana. The rent was controlled such that they cannot increase it except under stringent conditions. However, the public rent control administration resulted in the loss of investment with an adverse effect on the rental housing stock. Asabere (2007), in his conclusion, suggests several alternatives to improve the conditions of the housing sector through privatization. This scenario suggests that affordability in the housing sector remains an issue.

In breaking with the norm of looking at issues solely in Ghana, Grant (2007) looked at the effect of global money on the housing market in Ghana and, more specifically, in the upper-income neighborhoods in Accra. Grant "shed light on the international context of new houses in Accra" (Grant, 2007: 33). This internationalization was helped by the easier transfer of funds from abroad to Ghana, but also coincided with easier access to information through the internet. Nevertheless, most people buying units in these new developments were locals, followed by returnees, then non-resident Ghanaians. This study showed some of the relationships between Foreign Direct Investment and the local housing market in Ghana.

Two articles were written about Malawian real estate issues in 2007. The article by Nyasulu and Cloete (2007) discussed the unaffordability of housing in the urban areas of Malawi, along with the limited financial offerings to assist the urban population to access housing. Meanwhile, Manda (2007) writes about the efforts of an NGO and its involvement in helping urban poor raise funds to access land for building. The theme of housing for the poor and its finance is a critical issue in sub-Saharan African urban areas.

Tomlinson (2007) discusses the challenges banks face with lending to the black population of South Africa; the government has been in negotiations with the banking sector to expand its lending base. The paper concludes that the poles had shifted in many cases; by the time the negotiations had taken place, the house prices had risen above their affordability levels; hence, the process is stalled. So, the question remains who benefited from the negotiations? This question arises because the entire process of negotiation becomes futile at the expense of the prospective homeowner.

One group that should have benefited from these negotiations were the black women of South Africa. Ndinda and Uzodike (2008) discuss access to housing finance for women. Their research showed that despite the constitutional provisions for women and housing rights since 1994, many organizations that were geared towards assisting women towards homeownership still had a long way to go in achieving equality in terms of access to housing finance. This work shows that housing finance for the poor is still one of the very many challenges in the sub-Saharan African continent.

The slums of Nairobi have gained considerable worldwide recognition as some of the densest slums, and they offer housing for a large percentage of the city's population. Gulyani and Talukdar (2008) show that slums are expensive rental alternative compared to the amenities they offer. Despite the high rentals, there is no investment in improving the area or constant amenities available to the tenants. Hence, the debate continues whether there is a need to regulate or control the rents and whether the landlords are villains or entrepreneurs. Slums are in many African cities, and it is essential to have conversations on how they fit into the real estate market in cities and how they are viewed and dealt with from a regulation perspective.

Udechukwu (2008) goes through the process of identifying the obstacles that exist about individual homeownership in Nigeria. He lists all the land issues and financial issues such as availability and affordability of finance. He considers the different financial housing policies. Udechukwu also discusses the affordability from the demand perspective as well by looking at building material and prices. As in many African countries, the informal sector comes into play as it is the primary support to the housing delivery process in Nigeria as in the rest of the continent.

In 2009, Boamah published an article on housing and real estate in Ghana. It was interesting because, for the first time, a study was done on the possibility of Secondary Mortgage Markets (SMM) being the appropriate way to finance the housing market in Ghana (Boamah, 2009). The author concluded that SMM is not the best option for Ghana. The country first needed to solve several other problems, including land titling and adjudication procedures before venturing fully into SMM. It is interesting to start testing African markets for different financial housing solutions.

Obeng-Odoom (2010) published his study on different ways of transferring and remitting funds from Australia to Ghana and other remittances worldwide and their effect on urban real estate and homebuilding. He highlighted the problems with acquiring land and the inflation in Ghana. The remitting process was relatively smoother than other international places, but the people had to work more jobs to be able to send money to Ghana for the building of their private houses. Obeng-Odoom found that on average, Ghanaians living in Australia finished their houses faster than Ghanaians living in the UK. This study highlighted the importance of remittances to the housing sector in Ghana.

Informal settlements and their potential of accumulating capital and wealth for their owners was a topic of interest for Mooya and Cloete (2010). This is one of the first studies to look at informal settlements as real estate and their potentials. They conclude that introducing possible ways of harnessing the flexibility of informal rights to properties for institutional arrangements would encourage more activity in such markets, hence helping in the accumulation of capital for the urban poor.

Even though the private sector essentially controls the housing market in Nigeria, there are still significant problems with the housing supply. Nubi (2010), in his research on sustainable housing finance for Nigeria, concludes that it is important to start looking at the amalgamation of the construction companies, as none of them has the capacity of building large projects capable of denting the demand for housing. He also concludes that the government needs to start regulating the financial markets for the latent pension funds to be invested in housing. The article is a step closer to what needs to be done to get the housing markets to meet the demand for housing for the teeming population.

Gyamfi-Yeboah and Ziobrowski (2010: 339) attempted to test the effect of "capital and mortgage markets as an important step in moving emerging countries toward full economic development". The evidence showed that after the secondary mortgage market was introduced in 2001, not much change had happened as they were already integrated into the system. The economic conditions had not changed much during the same period.

There is no doubt that data availability is crucial in real estate studies, and trends with historical data are very important for the overall understanding of the markets. Das *et al.* (2011) used quarterly data from 1969 to 2009 to look for bubbles in the South African housing market and examine why there is "spillovers on South African consumption, using well-known econometric techniques" (Das *et al.*, 2011: 88). We have not seen such advanced use of econometric models in the African context except rarely.

Babawale and Adewunmi (2011) studied the impact of neighborhood churches on house prices and the negative externalities that they cause. The study concludes that the neighborhood churches negatively affect the housing prices in the area, but positive externalities would neutralize the housing prices. This study was conducted to assist in disputes over churches that come to court.

Gunter and Scheepers (2012) emphasize the issue of informal housing being a dead capital hindering their owners from moving up the residential ladder. Using a slum development in Johannesburg, they look at the restrictions on the informal housing developments and how it reduces the possibility of poverty eradication. They found that there is movement in the lower end of the market involving slum dwellers showing that it is not "dead capital", and it is important to embrace the neo-liberal principles that are assisting in this shift to achieve wealth creation for the poor. Interestingly, informal/slum housing is starting to attract more research in housing real estate.

Adeboye *et al.* (2012) discusses the financing of residential real estate in Nigeria. They deduce that the National Housing Funds (NHF) do not cater to the high-income earners, nor does it cater to low-income, non-salaried earners. These are drawbacks that need to be overcome in designing new long-term housing finance, more specifically when designing a scheme for the low-income, non-salaried earners. They identified some reasons for the lack of funding and suggest new sources for housing funding other than the traditional sources. They also suggest new regulatory measures for the financial sector to play a more active and stronger developmental role.

Anim-Odame (2012) discusses the residential real estate investment market and determines that the Ghanaian market is a good investment portfolio diversifier. The research supports the opinion that residential real estate markets play "a significant role in its own right in the national economy and also as a competitive investment asset class" (Anim-Odame, 2012: 302).

Many cities in developing countries had informal settlements sprout among their residential or edge areas post-independence due to a multiplicity of reasons, mainly a deficit of reasonably priced housing in the urban areas or the capital city. Nyametso (2012) looked at the effect of security of tenure in Accra and found that access to land security led to improvements in housing and living conditions for the urban poor.

Research conducted in Lagos, Nigeria, proved contrary to the hypothesis of rent theory, which states that the "distance travelled and cost of getting there is not a significant factor that affects property value" (Oloke *et al.*, 2013: 642). Interestingly, other factors, such as number of rooms, good road drainage and security, were leading factors that affect property prices in Magodo Neighborhood in Lagos, Nigeria.

Gunter (2014) conducted research on the shack rental market. The qualitative research yielded some interesting results and attributed the rise in the shack rental market to the failure of the market to absorb these low earners in the homeownership market. The author also realized that landlords consider their actions in renting out shacks as a noble cause, allowing them to help the urban poor find shelter in the city. The research asserts the importance of the shack rental market in addressing the shortages in the housing market.

Hernando De Soto has remained a controversial figure for over 30 years in the field of housing in developing countries (De Soto, 2001a, 2001b; Geyer and Geyer, 2014). Geyer and Geyer (2014) highlight how both sides of the arguments have missed some very important aspects that are crucial for the poor and especially in South Africa. The argument stems from the different perceptions to "the transformation of housing and business regimes in developing countries from informal to formal as a potentially powerful mechanism to turn large amounts of latent capital into life, working capital" (*ibid*:35). The paper argued, among other things, that a title deed is not a guarantee for access to credit. Their research led them to conclude that the vibrancy of the informal markets depends on whether they are registered or not.

By 2012, almost 500 000 state-owned housing units built during the apartheid era were privatized. This development triggered support and opposition in the housing sector (Marais *et al.*, 2014). The Hernando De Soto supporters claimed that it would release "dead capital" onto the market. The proponents or the "neo-Marxists" predicted landlordism and displacement of the poor, which did not happen. Marais *et al.* (2014) realized that historical factors, such as blacks not having rights to own property during the apartheid era, led to people valuing their housing and not willing to sell their units. They neither released the "dead capital", nor did they become landlords.

Irumba (2015) asserts that despite having a leasehold or freehold on a house in Kampala (Uganda), the property's value is enhanced by structural attributes such as parks, gardens, pools, and maid's quarters within the vicinity of the property. Although, the market values public leaseholds higher in comparison to freehold based on charges and land premium paid to the government for leasehold, thereby giving developers and buyers confidence (*ibid*). However, Matongela (2015) argues that land supply as a variable is the most substantial determinant of house prices in Namibia amidst other variables identified by Irumba (2015). Matongela (2015) advocates for the promotion of sites and services by the government to ease access to land towards interested house owners, regulate house prices, and regulate ill practices by speculators holding on to land without developing it.

Sani and Gbadegesin (2015) conducted a study in Kaduna to determine the causative factors of rental housing defaults in Nigeria based on the constant rifts between landlords and tenants. The study shows that while tenants are selected using different yardsticks and different rental tenures, the landlord-tenant relationship is made soar because the tenancy agreements are usually bridged, and many tenants point to landlords' harsh treatment and policy (*ibid*). Hence, they called for effective regulation to refine the residential property management schemes.

Kuma (2015) highlights the importance of housing finance access towards homeownership and the hindrance it has caused interested homeowners in northern Nigeria. He states that there are four mortgage firms in the regions and the conditions are not favorable or affordable to the majority. Hence, interested homeowners have resulted in other means of accessing housing finance other than mortgages, leading to incremental forms of building their houses (*ibid*).

Akinlabi (2015) argues that multipurpose cooperative societies as informal financial institution have the natural tendency to significantly support housing production as against government funding all housing projects and providing funds to formal financial institutions. He noted that while these societies are willing to support their members, they are grossly incapacitated financially, and many cooperative societies are fraught with financial misdemeanor.

Teye *et al.* (2015) highlight the demand and supply constraint inhibiting formal mortgage market and institutions towards housing production in Ghana. These constraints are the financial institution limited capital, inability to identify genuine credit worthy potential borrowers, high interest rates, expensive nature of houses due to the unfavorable economic situation at the expense of low-income earners, and attitudinal culture to loans (*ibid*). The authors emphasize the need to develop "a well-functioning housing finance system" to boost Ghana's real estate mortgage market.

Adjekophori and Adebiyi (2016) highlight the importance of the capital market in economic development and the need to annex redundant funds from the capital market to support housing production and economic development as a product of increased housing finance. Ajayi *et al.* (2016) argue that while there are several challenges impeding housing affordability and delivery in Nigeria, access to buildable land and an ill-managed mortgage system are the core challenges. Other challenges are high interest rate for mortgages, issues relating to the 1978 Land Use Act, high cost of perfecting title documents, and statutory and professional design fees.

Another similar study was carried out in Nigeria by Abidoye and Chan (2016) through the lens of estate valuers to identify variables responsible for the value of residential properties. These variables being ranked in the order of importance reflect the property's location, the surroundings and physical environment, physical stability and building durability, level of neighborhood security, and how long the property has existed (*ibid*).

Kibunyi *et al.* (2017) carry out a study to determine variables that influence house prices in Kenya. Their study reveals that housing prices are primarily affected by economic indices such as "GDP, diaspora remittances, lending rates, loans to real estate, and construction cost" to a varying degree. House prices do not do well with inflation, while a stable economy around the GDP and stock exchange market prolong stability in house prices (*ibid*).

Owusu-Ansah *et al.* (2017) discuss the effect of temporal aggregation for price indices in upcoming real estate market through a study in Ghana. The findings show that the timeline of temporal aggregation makes no difference in price indices, but lower temporal aggregation has more impact on house prices indices in comparison to broader levels.

Bondinuba *et al.* (2017) explore the factors and conditions that motivate or demotivate microfinance institutions within the low-income housing market and the resultant effect on the housing sector in Ghana. The motivating factors include but are not limited to the need for share of the market size, the magnitude and demand for home ownership, and the propensity for growth. However, many other factors discourage investment by these institutions, such as high interest rate, high cost of land, high cost of construction, and expensive building materials. They noted that demotivators for microfinance institutions exceed what can be considered as motivators, and as such are having a negative impact on the low-income housing market.

Nkechi *et al.* (2018) conducted a study across Nigeria, Ghana, and Zambia to measure the impact of housing finance on economic growth and development. While there is evidence of contribution of housing finance in Nigeria through mortgages and commercial banks to economic growth, the effect is insignificant in Ghana and almost non-existent in Zambia.

The Ghana case is attributed to house acquisition through self-help and incremental basis which takes a lot of years while the situation is Zambia is exacerbated because of mass poverty. Invariably, there is an economic undergrowth in the housing finance market in West African region which is also a reflection of the economic situation, Adu-Gyamfi, et al. (2020).

Following the volatility in Namibia's real estate prices and bit to justify the presence of a bubble and its determinants, Nyambe *et al.* (2018) conducted a study from 2007 to 2014, covering the period after the 2008 global financial crisis. Their study confirmed the bubbles and show that there was "one major bubble, two intermediate bubbles and eighteen minor bubbles" within that period (*ibid*:93). While the minor bubbles have no threats attached to the periods, the intermediate and the major bubbles call for serious concerns and actions to be taken. The period of the major bubble coincides with first quarter of 2014, and it was attributed to the non-availability of serviced land for property development and low interest rates, which led to the flourishing of the black land market (*ibid*). This study alludes to Matongela (2015)'s finding attributing serviced land supply as a determinant affecting prices of properties.

Ikekpeazu (2018) discusses the need for the government to adopt policies that will favor the housing sector and mortgage finance in Nigeria by strengthening institutions involved in housing production. He highlights the prominent role that the government occupies in housing delivery in addressing pertinent issues such as "the dearth of long-term financial instruments, absence of mortgages liquidity, weak capital base, inadequate primary mortgage institutions, weak corporate governance, inadequate skilled labour and high cost of building materials" (*ibid*:1).

Akaabre *et al.* (2018) discuss the growth of the informal rental housing market in Ghana based on inadequate housing with the growth worsened due to the rapid urbanization and socio-economic situation of the citizens. Apart from job creation, the effort of these informal agents is meeting the needs of the urban poor, poor immigrants, and the landlords making their properties available (*ibid*). However, there is a need for the regulation for the informal rental housing market to protect the public in their bit to serve the gap not met by the formal market.

Kaulihowa and Kamati (2019), in their study, established the existence of housing price volatility in Namibia and that mortgage loans and GDP as a macroeconomic variable constitute the highest determinant of price volatility. Hence, the need for government to consider the multiplier effects of macroeconomic variables on economic growth and house acquisition within the housing policy formulation.

Alkali *et al.* (2020) discuss how negative information in the real estate housing market can influence prices volatility in Abuja. Their conclusion reveals that negative information will depend on typology of the housing property, but two- and three-bedroom flats seem to be the most affected due to higher demand and needs assessment.

Owusu-Manu *et al.* (2019) discuss the housing attributes of a property and how it affects housing price in relation to urban planning in Ghana. Amongst many attributes such as housing typology, age of house, finishes, floor and land area, the location of the building is ranked highest by the respondents. The findings of this study align with a previous study carried out by Abidoye and Chan (2016) in terms of variables that determine the price of a property. Hence, the enhancement of all neighborhoods with necessary infrastructure will have a resultant price effect on the buildings.

Gavu and Owusu-Ansah (2019) discuss the existence of residential submarket at different scales of aggregation in Ghana from the methodological point of view. Having proven the existence of the real estate submarket, they advocate for the use of different models from homeownership and rental markets to see the impact of dependent and independent variables.

Gavu *et al.* (2019) conducted a study to understand the market feel of residential rental value determinant in Ghana, exploring different variables across the different housing and rental typologies. The stakeholders' perception reflects that variables such as piped water and power supply, number of rooms, housing condition, and typology are the critical determinants of residential rental values. Other determinants are a well-landscaped environment, closeness to worship centres, and recreational facilities.

Ampofo (2020) conducted a study in Ghana on mortgage administration across mortgage institutions. The result shows mortgage repayment is affected by fluctuation in exchange rates, high interest rates, and high premium paid for expensive properties. Mortgagees mostly use fixed-rate method, among other repayments, to access housing finance.

Embaye *et al.* (2021) conducted a study using machine learning methods to predict the rental value of different households across Tanzania, Uganda, and Malawi. Beyond external indices and factors that could influence the rental value of a house and the welfare of the residents, they placed value on the components and services in and around the house as variables in the machine methods for predicting the rental value (*ibid*). The machine methods used are OLS, Ridge, LASSO, Tree, Bagging, Forest, and Bosting.

Conclusion

The literature review suggests that academic interest in African property markets is growing and widening in scope and geography. Research in the housing market has tackled several sub-themes, such as affordability of housing, housing policy objective, determinants of property values, housing finance, valuation, and the characteristics of formal and informal markets.

The existing literature emphasizes the institutional complexities associated with residential markets and reflects on their gradual maturity. In recent years there has been a growing interest in property finance and research that is more of a quantitative nature. The type and scope of research in Africa continue to be constrained by a lack of data.

Much of the focus of the literature on Africa has been on Nigeria, with authors also focusing on South Africa and Ghana. There is also evidence that research in several other countries across the continent is receiving growing attention. The growing maturity of the sub – Saharan property market means that the body of knowledge reviewed in this chapter will evolve. While specific topics such as the development of appropriate housing policies are expected to remain central to the research agenda, the focus is expected to turn to the role of financial markets in furthering property ownership across the continent. Also, there is a need for more comparative studies on the core topical issues to be carried out by academics across Africa along global concerns.

References

Abidoye, R. B., & Chan, A. P. (2016). Critical determinants of residential property value: Professionals' perspective. *Journal of Facilities Management, 14*(3), 283–300.

Adeboye, A., Rod, G., Felix, H., & Olomolaiye, P. (2012). An overview of residential real estate financing in Nigeria. *Journal of Real Estate Literature*, *20*(2), 263–288.
Adjekophori, B., & Adebiyi, S. O. (2016). Effectiveness of capital market derivatives in housing delivery of Nigeria emerging market. *Studies and scientific researches. Economics Edition*, *23*, 36–53.
Adu-Gyamfi, A., Cobbinah, P. B., Gaisie, E., & Kpodo, D. D. (2020, September). Accessing private rental housing in the absence of housing information in Ghana. In *Urban Forum* (pp. 1–19). Springer, Netherlands.
Ajayi, O., Ajayi, O., Akinsiku, O., & Osunsanmi, T. (2016). Strategies for housing affordability in Nigeria. *Journal of Construction Project Management and Innovation*, *6*(Sup-1), 1620–1632.
Akaabre, P. B., Poku-Boansi, M., & Adarkwa, K. K. (2018). The growing activities of informal rental agents in the urban housing market of Kumasi, Ghana. *Cities*, *83*, 34–43.
Akinlabi, A. J. (2015). Informal financial institutions and sustainable housing finance in southwestern Nigeria. *The Business & Management Review*, 7(1), 95–104.
Akpom, U. N. (1996). Housing attributes and the cost of private rental buildings in Lagos, Nigeria: A hedonic price analysis. *The Review of Regional Studies*, *26*(3), 351–365.
Alkali, M., Sipan, I., & Razali, M. N. (2020). The effect of negative information on the volatility of real estate residential prices in Abuja, Nigeria. *International Journal of Housing Markets and Analysis*, *13*(2), 267–280.
Ampofo, J. A. (2020). The nature of mortgage repayment plans in Ghana. *Finance & Accounting Research Journal*, *2*(3), 91–104.
Anim-Odame, W. (2012). The nascent residential investment market in Ghana. *Journal of Real Estate Literature*, *20*(2), 289–303.
Arimah, B. C. (1992). Hedonic prices and the demand for housing attributes in a third world city: The case of Ibadan Nigeria. *Urban Studies*, *29*(5), 639–651.
Arimah, B. C., & Adinnu, F. I. (1995). Market segmentation and the impact of landfills on residential property values: Empirical evidence from an African city. *Journal of Housing and the Built Environment*, *10*(2), 157–171.
Asabere, P. (2007). The demise of the rent-controlled public housing programmes of Ghana: The story of the Tema development corporation. *Urban Studies*, *44*(10), 1919–1935.
Babawale, G. K., & Adewunmi, Y. (2011). The impact of neighbourhood churches on house prices. *Journal of Sustainable Development*, *4*(1), 246.
Boamah, N. A. (2009). Secondary mortgage market (SMM): Is it right for financing housing in Ghana? *Journal of Science and Technology (Ghana)*, *29*(1), 17–27.
Bondinuba, F. K., Hedidor, D., Opoku, A., & Teye, A. L. (2017). De/motivations in housing microfinance delivery in Ghana. *Property Management*, *35*(5), 528–544.
Clauretie, T., & Daneshvary, N. (1993). A note on the ranking of real estate authors: where else do they publish and who cares? *Journal of Real Estate Research*, *8*(3), 445–453.
Das, S., Gupta, R., & Kanda, P. (2011). International articles: Bubbles in South African house prices and their impact on consumption. *Journal of Real Estate Literature*, *19*(1), 69–91.
De Soto, H. (2001a). The mystery of capital. *Finance and Development*, *38*(1), 29–33.
De Soto, H. (2001b). Dead capital and the poor. *Sais Review*, *21*(1), 13–43.
Diaz III, J., Black, R. T., & Rabianski J. (1996). A note on ranking real estate journals. *Real Estate Economics*, *24*(4), 551–563.
Embaye, W. T., Zereyesus, Y. A., & Chen, B. (2021). Predicting the rental value of houses in household surveys in Tanzania, Uganda and Malawi: Evaluations of hedonic pricing and machine learning approaches. *PLoS One*, *16*(2), e02449531.
Ezinwanne Udechukwu, C. (2008). Obstacles to individual home ownership in Nigeria. *International Journal of Housing Markets and Analysis*, *1*(2), 182–194.
Gavu, E. K., Gruehn, D., Schulte, K. W., & Asante, L. A. (2019). Stakeholders' perception of residential rental value determinants in Ghana. *Journal of African Real Estate Research*, *4*(1), 42–70.
Gavu, E. K., & Owusu-Ansah, A. (2019). Empirical analysis of residential submarket conceptualisation in Ghana. *International Journal of Housing Markets and Analysis*, *12*(4), 763–787.

Geyer, H. S., & Geyer Jr, H. S. (2014, March). Residential transformation in South Africa – reopening the "dead" capital debate. *Urban Forum, 25*(1), 35–55.

Gibler, K. M., & Ziobrowski, A. J. (2002). Authors' perceptions and preferences among real estate journals. *Real Estate Economics, 30*(1), 137–157.

Grant, R. (2005). The emergence of gated communities in a West African context: Evidence from Greater Accra, Ghana. *Urban Geography, 26*(8), 661–683.

Grant, R. (2007). Geographies of investment: How do the wealthy build new houses in Accra, Ghana? *Urban Forum, 18*(1), 31–59.

Gulyani, S., & Talukdar, D. (2008). Slum real estate: The low-quality high-price puzzle in Nairobi's slum rental market and its implications for theory and practice. *World Development, 36*(10), 1916–1937.

Gunter, A. (2014). Renting shacks: Landlords and tenants in the informal. *Urbani izziv, 25*, S96–S107.

Gunter, A., & Scheepers, L. (2012, June). "Crisylida capital": Hatching informal township property markets to benefit low-income residents in Johannesburg, South Africa. *Urban Forum, 23*(2), 165–180.

Gyamfi-Yeboah, F., & Ziobrowski, A. J. (2010). The integration of mortgage and capital markets in emerging economies – evidence from South Africa. *The Journal of Real Estate Finance and Economics, 41*(3), 339–353.

Hardin III, W. G., Liano, K., & Chan, K. C., (2006). Influential journals, institutions, and researchers in real estate. *Real Estate Economics, 34*(3), 457–478.

Hoogendoorn, G., Mellett, R., & Visser, G. (2005). Second homes tourism in Africa: Reflections on the South African experience. *Urban forum, 16*(2), 112–154.

Ikekpeazu, O. F. (2018). Mortgage financing for housing delivery in Nigeria: The governmental role in development of an effective mortgage market. *PM World Journal*, 7(1), 1–18.

Irumba, R. (2015). An empirical examination of the effects of land tenure on housing values in Kampala, Uganda. *International Journal of Housing Markets and Analysis, 8*(3), 359–374.

Kajimo-Shakantu, K., & Evans, K. (2006). The role of banks in the provision of low-income housing finance in South Africa: Can they play a different role? *International Journal of Strategic Property Management, 10*(1), 23–38.

Kamete, A. Y. (1997) Formal private sector housing finance institutions and low-income groups in Zimbabwe: Peculiar requirements, operations and realities. *Journal of Social Development in Africa. 12*(1), 83–91.

Karley, N. K. (2003). Challenges in mortgage lending for the underserved in South Africa. *Housing Finance International, 18*(1), 27–33.

Kaulihowa, T., & Kamati, K. (2019). Determinants of house price volatility in Namibia. *International Journal of Housing Markets and Analysis, 12*(4), 807–823.

Kibunyi, D., Ndiritu, S. W., Carcel, H., & Gil-Alana, L. A. (2017). Real estate prices in Kenya: is there a bubble? *Journal of Housing and the Built Environment, 32*(4), 787–804.

Knight, J. R., Herrin, W. E., & Balihuta, A. M. (2004). Housing prices and maturing real estate markets: Evidence from Uganda. *The Journal of Real Estate Finance and Economics, 28*(1), 5–18.

Kuma, S. S. (2015). Assessing the challenges of access to housing finance in the North Central States of Nigeria. *Ethiopian Journal of Environmental Studies and Management, 8*(2), 161–170.

Lemon, A., & Clifford, D. (2005). Post-apartheid transition in a small South African town: interracial property transfer in Margate, KwaZulu-Natal. *Urban Studies, 42*(1), 7–30.

Manda, M. A. Z. (2007). Mchenga – urban poor housing fund in Malawi. *Environment and Urbanization Urbanisation, 19*(2), 337–359.

Marais, L., Sefika, M., Ntema, J., Venter, A., & Cloete, J. (2014,). Towards an understanding the outcomes of housing privatisation in South Africa. *Urban Forum, 25*(1), 57–68.

Matongela, A. M. (2015). A quantitative investigation into determinants of house prices in Namibia. *Journal of Economics and Sustainable Development, 6*(4), 1–7.

Megbolugbe, I. F. (1989). A hedonic index model: The housing market of Jos, Nigeria. *Urban Studies, 26*, 486–494.

Merrill, S. R. (2001). Low-and moderate-income housing finance in South Africa: Making progress in a troubled environment. *Housing Finance International, 15*(3), 51–64.

Mooya, M. M., & Cloete, C. E. (2010). Property rights, real estate markets and poverty alleviation in Namibia's urban low income settlements. *Habitat International, 34*(4), 436–445.

Moss, V. (2001). The state of housing finance in South Africa. *Housing Finance International, 16*(2), 30–35.

Ndinda, C., & Uzodike, U. (2008). Accessing housing finance in South Africa: The role of women activism. *Wagadu, 6*, 75–91.

Nkechi, O. T., Samuel, A. O., & Ifurueze, M. S. (2018). Housing finance market and economic growth of west Africa region: A study of Nigeria, Ghana and Gambia. *The Business & Management Review, 9*(3), 188–198.

Nubi, T. G. (2010). Towards a sustainable housing finance in Nigeria: The challenges of developing adequate housing stock and a road map. *Housing Finance International (Online), 24*(4), 22.

Nyambe, C., Vijayakumar, K. R., & Gnitchogna, R. (2018). Are there bubbles in the housing market of Namibia? An analysis of historical evidence. *International Science and Technology Journal of Namibia, 9*, 85–95.

Nyametso, J. K. (2012, September). Resettlement of slum dwellers, land tenure security and improved housing, living and environmental conditions at Madina Estate, Accra, Ghana. *Urban forum, 23*(3), 343–365.

Nyasulu, E. C. & Cloete, C. E. (2007). Lack of affordable housing finance in Malawi's major urban areas. *Property Management, 25*(1), 54–67.

Obeng-Odoom, F. (2010). Urban real estate in Ghana: A study of housing-related remittances from Australia. *Housing Studies, 25*(3), 357–373.

Ogu, V. I., & Ogbuozobe, J. E. (2001). Housing policy in Nigeria: towards enablement of private housing development. *Habitat International, 25*(4), 473–492.

Okonkwo, O. (1998). Housing finance and housing delivery systems in Kenya: Bottlenecks, recent developments and the way forward. *Housing Finance International, 12*, 14–26.

Oloke, C. O., Simon, F. R., & Adesulu, A. F. (2013). An examination of the factors affecting residential property values in Magodo neighbourhood, Lagos State. *International Journal of Economy, Management and Social Sciences, 2*(8), 639–643.

Onibokun, G. A. (1971). Housing finance in Nigeria. *Town Planning Review, 42*(3), 277–292.

Owusu-Ansah, A., Adolwine, W. M., & Yeboah, E. (2017). Construction of real estate price indices for developing housing markets. *International Journal of Housing Markets and Analysis, 10*(3), 371–383.

Owusu-Manu, D. G., Edwards, D. J., Donkor-Hyiaman, K. A., Asiedu, R. O., Hosseini, M. R., & Obiri-Yeboah, E. (2019). Housing attributes and relative house prices in Ghana. *International Journal of Building Pathology and Adaptation, 37*(5), 733–746.

Quayson, A. (2007). Ghana primary mortgage market initiative. *Housing Finance International, 22*(2), 52.

Rakodi, C. (1995). Housing finance for lower income urban households in Zimbabwe. *Housing Studies, 10*(2), 199–228.

Redman, A. L., Manakyan, H., & Tanner, J. R. (1998). The ranking of real estate journals: A citation analysis approach. *Financial Practice and Education, 8*(2), 59–69.

Redman, A. L., Manakyan, H., & Tanner, J. R. (1999). A normalized citation analysis of real estate journals. *Real Estate Economics, 27*(1), 169–182.

Sani, K. S., & Gbadegesin, J. T. (2015). A study of private rental housing market in Kaduna Metropolis, Nigeria. *International Journal of Humanities and Social Science, 5*(8), 173–183.

Simons, R. A., & Karam, A. (2012). Professional real estate activities and academic journal importance among African scholars: A snapshot of the 2008 African real estate society meeting. *Journal of African Real Estate Research, 2*(1), 2–17.

Smets, P. (1996). Community-based finance systems and their potential for urban self-help in a new South Africa. *Development Southern Africa, 13*(2), 173–187.

Teye, J. K., Teye, I., & Asiedu, M. O. (2015). Financing housing in Ghana: Challenges to the development of formal mortgage system. *Journal of Housing and the Built Environment, 30*(1), 1–16.

Tomlinson, M. R. (1999). Ensuring mortgage repayment in South Africa. *Housing Finance International, 13*, 40–43.

Tomlinson, M. R. (2002). Efforts and errors: South Africa's search to extend housing finance to low-income households. *Housing Finance International, 17*(2), 8–14.

Tomlinson, M. R. (2007). The development of a low-income housing finance sector in South Africa: Have we finally found a way forward? *Habitat International, 31*(1), 77–86.

Ubogu, R. E. (1988). Demand for mortgage finance in Nigeria: A cross-sectional analysis. *Savings and Development, 12*(3), 271–286.

Ukoha, O. M., & Beamish, J. O. (1997). Assessment of residents' residents' satisfaction with public housing in Abuja, Nigeria. *Habitat International, 21*(4), 445–460.

UN-Habitat. (1996). *Istanbul Declaration-United Nations Conference on Human Settlements.* Instanbul (Turkey). United Nations.

UN-Habitat. (2016). The New Urban Agenda. The United Nations General Assembly, as adopted at the United Nations Conference on Housing and Sustainable Urban Development (Habitat III) in Quito, Ecuador in October 2016 (Quito).

United Nations. (2015). *Transforming Our World, the 2030 Agenda for Sustainable Development.* United Nations.

Webb, J. R., & Albert, J. D. (1995). Evaluating the real estate journals: The mainstream finance prospective. *Journal of Real Estate Research, 10*(2), 217–226.

Weeks, H. S., Finch, J. H., & Hardin III, W. G. (2007). Endowed real estate positions and the faculty who hold them. *Journal of Real Estate Practice and Education, 10*(1), 61–79.

10

INCLUSIVE, AFFORDABLE, AND SMART HOUSING IN AFRICA

Prisca Simbanegavi and Kolawole Ijasan

Introduction

Developing inclusive, affordable, and smart housing is one of the challenges faced by many countries particularly in Africa, Latin America, and Asia who aim to achieve the UN Sustainable Development Goals (10 and 11) (United Nations, 2019). It is important to know what is required in building good quality housing assets which maintain investment value in the future. In many African countries, discourse on social and inclusive housing has had its central thesis on righting the wrongs of the colonialism and apartheid exclusion systems. This chapter makes an argument that it is not enough to focus on just providing affordable housing if there are no well thought-out provisions for financial sustainability in the houses we build. An over concentration of financially disadvantaged people cannot yield a functional housing market which is viable and sustainable. This chapter will put guidelines for building inclusive, affordable, eco-friendly, and smart housing in Africa.

There are four sections in this chapter:

A. Housing Shortage Under Colonialism and Apartheid
B. Inclusive Housing
C. De-risking Inclusive, Eco-friendly, Smart Affordable Housing
D. Guidelines for Future Inclusive Housing
E. Discussion Questions and Areas of Further Research

A. Housing Shortage Under Colonialism and Apartheid

Extant literature shows how pieces of legislation were used to orchestrate land tenure systems that deprived and disadvantaged the natives in Africa (Becker, 2020; Thompson, 2001). Of particular interest is South Africa's Native Land Act 27 of 1913 which classified all people into racial group areas. Through this act, Africans were banned from owning land as it was reserved for the white race (known as 'territories' or colonies) while the black race was consigned to rural lands, which were called 'homelands'. Over time, African labourers were required for

DOI: 10.1201/9780429279256-10

mining and farming in these territories. This phenomenon made the rural-urban income differentials attractive and forced black South Africans to move into the territories in search of better employment opportunities. However, this deprivation created a major housing shortage in the territories. Later, in the early 1920s, the Housing Act 32 of 1920 and the Natives Urban Areas Act 35 (1923) created living spaces for all racial groups in the urban areas. After the 1948 election, the National Party, following the example of the British and the Trek Boers before them, decided to severely curb the movement of Africans in the urban regions. Subsequently, the Group Areas Act 41 (1950) physically separated people into different racial residential areas. Thus, through the Group Areas Act and the Population Registration Act of 1950, land became the restricted possession of the white racial group. Property transactions were not permitted for Africans (natives) who were, from thereon, deemed 'disqualified persons' (Becker, 2020). Without the ability to own or rent property, informal settlements sprung up as the only accommodation available to Africans from the homelands. This segregation of the past has led to discriminatory policies and the misallocation of resources in the housing sector causing huge housing shortages in Africa. Approximately, 20% of urban households reside in informal settlements in South Africa despite decades of comprehensive government subsidised housing delivery (Simbanegavi, 2021).

Extant literature shows the un-dwindling legacy of apartheid that created 'two incompatible worlds' where people live side by side due to the separationist city planning policies (Charlton, 2004). Many people live in informal settlements, a prototype solution to the urban housing crisis. There exist a substantial number of people who are homeless living in abject poverty. In contrast to this sprung the concept of rich 'gated communities' in an effort to counter high levels of crime in neighbourhoods. The gated communities are security estates that are heavily guarded where people who live there have access to top notch amenities such as golf and gym facilities. Informal settlements usually exist with barely nothing to call a place a home (see Figure 10.1). The figure shows some informal settlements where many people live without access to basic amenities and infrastructure.

It is clear how colonialism and apartheid shaped the distribution and ownership of land and housing throughout Africa before the attainment of democracy (Moss *et al.*, 2013). This is the manner in which the discriminatory policies ultimately distorted Africa's residential markets by racially segregating neighbourhoods. A photographer by the name Johnny Miller has been able to capture the two worlds juxtaposed on one location in most African cities. The most visited global city of Sandston in South Africa is just separated by the highway from a low income low income neighbourhood called Alexandra (Figure 10.2).

Shortage of affordable housing, specifically for black African working class people in South Africa, and is dire in Africa. Added to this is the influence of population increases in many countries, with Nigeria and China taking the lead in the increase (UN-HABITAT, 2014). Apartheid in South Africa made residential markets inefficient, creating a housing shortage of about 2.1 million houses (approximately 12.5 million people by 2011) (National Department of Human Settlements, 2016). Thompson (2001) clearly shows how under apartheid pieces of legislation were used to deprive black people in South Africa of land. Overall, the housing shortage in many African countries has persisted and worsened due to global financial crises, increasing the severity of disasters, political and religious unrest, migration, population increases, and increases in urbanisation rates. Projections by the World Bank (2018) indicate that the sub-Saharan Africa region leads other regions in urban population living in informal settlements (slums) from 1990 to 2030 and the trend is actually rising. This is compared to

FIGURE 10.1 A general view of Stjwetla informal settlement on the banks of the Jukskei River in Alexandra, Johannesburg, taken on 22 June 2020

Source: Webster, D and Haffejee, I Accessed: 28 July 2021 www.newframe.com/bringing-de-densification-home-in-alexandra/

other regions which include the Middle East and North Africa, Latin America and Caribbean, South Asia, and East Asia and Pacific where the numbers have been decreasing, with the lowest number in Europe and Central Asia as shown in Figure 10.3.

King *et al.* (2017) pointed out that millions of people around the world are homeless where approximately 880 million people live in slums in developing countries and 330 million households require urban housing. Informal settlements are an indication of the gravity of the housing shortage in some parts of the world. The world's largest slums are situated in Khayelitsha in Cape Town in South Africa with 400,000 people; Kibera in Nairobi Kenya with 700,000 people; Dharavi in Mumbai India with 1,000,000 people; Neza in Mexico with 1,200,000 people; and Orangi Town in Karachi, Pakistan, with 2,400,000 people (World Bank, 2018). These residential areas are characterised by sub-standard housing for low-income earners in abject poverty. This contrasts the 1976 Vancouver Declaration on Human Settlements that recognises housing as a basic human right and the Sustainable Development Goal (SDG) 11 of making cities and human settlements inclusive, safe, resilient, and sustainable (United Nation, 2019).

B. Inclusive Housing

First coined by Jupp *et al.* (1999), the concept of mixing people from diverse race and social economic status in inclusive housing takes two forms: 'silo' mixing involves households

FIGURE 10.2 Inequality in cities: Alexandra informal settlement and Sandton City

Source: Miller (2016) Accessed: 15 May 2018 https://www.behance.net/gallery/38385837/Kya-Sands-Bloubosrand

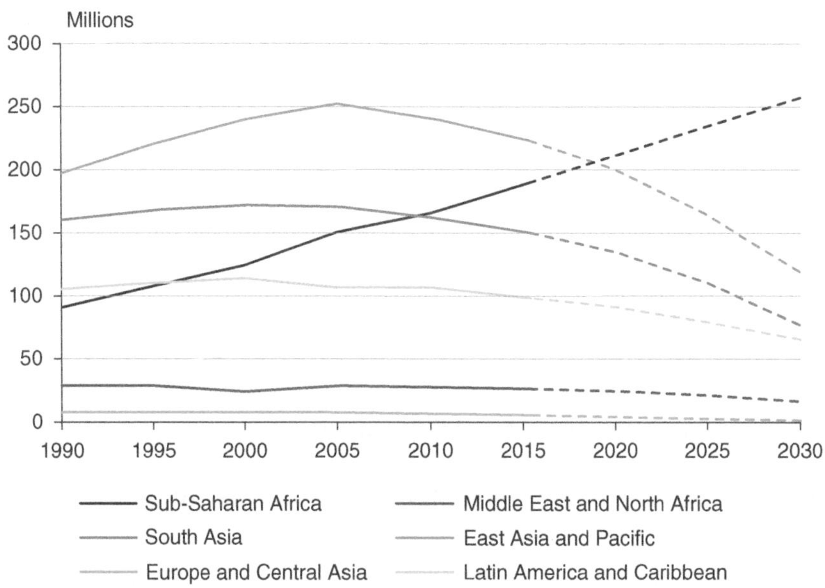

FIGURE 10.3 Population living in slums by region (1990 to 2030)

Source: World Bank, 2018

occupying designated topology; 'pepper potting' entails non-distinguishable mixing of low income households into increased density sustainable developments (Onatu, 2012; Osman *et al.*, 2011). The concept of inclusive housing is therefore, embedded in the efforts to reduce informal settlements, and to reverse the inequalities caused by colonialism. Indeed literatures shows a progressive shift in the way residential markets have developed globally, from distinctive low, medium, and high-income sub-markets towards inclusive housing. Inclusive housing is a mixture of the three sub-markets (social rental housing, subsidised mortgage housing, and the market rate housing) with different topologies in one development, which translates into mixing the poor and the rich (Khadduri and Martin, 1997). There is also evidence that better-off residents acted as role models who help poorer residents in finding better employment opportunities or raise expectations of attainment (Becker, 2020). Thus the argument is that households within an inclusive housing development should be local residents with a wide range of incomes and that such initiatives must be designed for those who fit into the surrounding neighbourhoods, who work and are already 'accepted' in the neighbourhood (van Gent and Musterd, 2013; Jupp *et al.*, 1999). Inclusive housing developments are placed in the hands of private developers for the development of mixed use neighbourhoods that have potential to integrate people of different classes. Arguably, this move may make government lose control of the delivery of social housing which will negatively affect the supply of housing for the lowest income people (Jones and Datta, 2010). Inclusive housing is highly preferred to social housing developments as they play an important role of poverty de-concentration and racial integration.

To promote inclusive housing, government's eagerness to join hands with the private sector is evidence for the role that private sector plays in building inclusive housing. However, despite the substantial investments made by government, inadequate housing still exists. While, housing policies encourage inclusive and affordable housing developments as a mixture of the three sub-markets private developers experience certain risks even in developed countries such as Australia, Netherlands, UK, and the USA (SAPOA, 2018; Hulse and Yates, 2017). Extant literature indicates that solving this problem requires reducing (de-risking) these risks for the private sector to make housing a good investment asset class that maintains value in the future. Massyn *et al.* (2015) emphasise the appropriate housing developments whose value is able to meet the costs incurred by the developers. Some aspects of reducing risks faced by housing developers include (i) applying the Public Private People Partnership Inclusive Housing Delivery Model (4Ps) of collaboration, (ii) using incentive based financing infrastructure, (iii) building inclusive housing in Greenfields, (iv) designing 'pepper potted' topologies in inclusive housing developments, (v) enable eco-friendly and smart affordable housing, and (vi) adopting private oriented asset and property management tools.

C. De-risking Inclusive and Affordable Housing

i. Applying a Public Private People Partnership Inclusive Housing Delivery Model (4Ps)

Through collaboration, the public and private sector and people community driven initiatives are able to build inclusive housing which minimises risks of failure (Eziyi, 2012). This is because inclusive housing developments become viable development projects rather than mere government driven social-spatial change programmes. Myeni and Mvuyana (2018) emphasised community participation as an important entry point to the political decision-making for initiating the negotiations needed for housing projects coordination and cooperation. The 4Ps collaboration can bring the much-needed funding. This is because raising funds through partnerships provides workable solutions (structured finance) towards housing where the public sector negotiates to exchange units of affordable housing for its financial contribution on land and bulk infrastructure through planning gain (Chigwenya, 2019). Graham (2016) emphasises that cost sharing with private sector infrastructure lowers risk and promotes stable long-term returns for inclusive affordable housing where institutional investors are drawn to invest in such residential developments. The private sector, and NGOs such as International Housing Solutions, Workforce Housing Fund, Clinton Foundation, and banks, to mention a few, may work with development partners to use city land for inclusive housing. To complete this value chain, there is need for ownership transparency through title deeds which is sometimes a neglected issue in social housing built for the poor in South Africa. Title deeds will link house owners to the mainstream mortgage market and general credit facilities to make housing affordable in South Africa.

In the 4Ps model, communities get involved in the delivery of housing within the inclusive concept. Governments have proved to be inefficient, slow, bureaucratic, and corrupt in providing quality housing developments that are investment assets (DFID, 2010). Similarly, the private sector has proven to be too self-serving when contracted for the provision of large-scale projects. However, when the two entities leverage their combined efforts, the resultant inclusive housing, such as Cornubia in Durban, South Africa, are better than the previous state subsidised housing constructed by government alone. The development managed to bring in other forms of real estate, such as retail, office, and industrial spaces. Moreover, there is evidence that community based (crowd-funding) approaches to housing enable delivery of housing that is appropriate through sweat equity and part contribution (Chigwenya, 2019; Montgomery *et al.*, 2018).

ii. Financing Infrastructure Through Incentive Based System

It is observed that government should use an incentive based system whereby the government pays for infrastructure (sewer, water, roads) which reduces capital requirements for private developers who develop such housing (Ferilli *et al.*, 2017). The incentive based financing system nullifies the need for compulsory inclusive policies as it removes shares risk to both government and private developers. Government and municipalities may essentially provide both land and bulk infrastructure in order to reduce the cost of these essential services to developers. This enables the private sector to build better housing that maintains investment value in the future and is inclusive enough to bring people closer to places of opportunity and employments. It is also important to investigate the appropriate location for inclusive housing in Africa.

iii. Building Inclusive Housing in Greenfields

The concept of Highest and Best Use (HBU) places inclusive housing in the outskirts of cities. This is consistent with the argument that affordability increases by facilitating outer suburban developments (McGreev, 2018). To correct this problem, Greenfields location should be accompanied by good infrastructure (roads, schools, hospitals, transport networks), and businesses to reduce investment risk. The argument is that Brownfields will eventually be 'built up' in the long term and so Greenfields is the default location for inclusive housing. This location may solve gentrification problems created when the poor are displaced (Bridge *et al.*, 2015). Government subsidy and grants may help the private sector build quality and bigger houses, especially for the low-income component of inclusive developments. Thus, the value of this government investment must be shared in the form of low-income housing units; otherwise, the total inclusive investment value contributed by government skews to the private sector.

iv. Designing 'Pepper Potted' Topologies in Inclusive Housing Developments

'Pepper potting' type of mixing is the ideal inclusive type of mixing whereby households get 'scattered' in a housing development plan regardless of race and income status, especially when inclusive housing is built with improved amenities, traffic, and security services (Taruvinga and Mooya, 2018). However, housing developments in African cities generally follow the 'silo' type where households are separated into low-, medium-, and high-income submarkets demarcated by roads within an inclusive housing development (Tunstall and Fenton, 2006). Open plans encourage people to walk to places of amenities increasing visibility and accessibility to housing, thereby reducing crime. This does not necessarily separate people by race but by income which broadly helps reduce areas of concentrated poverty. This way, households with differing income levels may upgrade and downgrade as their financial positions change. Inclusive policies should focus on mixing households according to income levels where different typologies rather than race, religion, and culture are designed for housing developments.

v. Build Eco-friendly and Smart Affordable Housing

Environmental degradation can be minimised when developers and communities endeavour to save the world through building eco-friendly and smart inclusive housing developments. An eco-friendly housing development should have less negative bearing on environment and climate than the conventional ones (Mohamad *et al.*, 2020). Housing can incorporate various Green Building Technologies (GBT) included in the list:

> Orienting the building to maximise natural daylight, minimize glare and heat radiation; natural ventilation; green roofs; solar energy; compact fluorescent lighting; cork flooring, using locally available or recycled material; minimizing heat radiation by landscaping; reducing water usage through low-flow fixtures; optimizing the building shape through thermal modelling; using energy efficient appliances and light fittings; optimising, upgrading or removing HVAC systems; lighting controls; harvesting rainwater techniques; using recycled greywater for flushing and having efficient drainage systems.
>
> *(Simbanegavi et al., 2019:4)*

Environment and climate rehabilitation is important and should not be weighed down by budget constraints, as the benefits are realisable in the long term. Extant literature find that improving energy efficiency in rented affordable housing units can contribute to making developments more affordable for residents in the long term (Chegut, 2016; Gruis and Nieboer, 2004).

The other important way of making inclusive housing affordable and supportive is through the installation of smart technologies (Mohamad *et al.*, 2020). As specified by World Health Organization (WHO), Covid 19 pandemic lockdown dictates that housing developments require access to Internet of Things (IoT) for housing to be inclusive. A smart house enables people to do many things that include online work from home away from offices (telecommuting), learn from home for students through software such as Microsoft Teams, Zoom, among others. The 'new normal' cuts commuting costs and enables people to save money and time when living in smart houses. Lastly, there is need for private sector oriented asset and property management strategies to ensure sustainability of the social housing topology within inclusive housing developments.

vi. Adopt Private Sector-Oriented Asset and Property Management Strategy

Given that inclusive housing has the social housing component in the development, there is need to use better asset and property management strategies for inclusive housing to be sustainable. This enables competitive resource allocation and income collection methods, without which a development might blight down into low-income housing. Asset management strategies increase the supply of quality housing stock while the property management tools maintain the stock, especially of the social housing component of an inclusive housing development. It is better for inclusive developers to source government land for inclusive housing as this makes such projects viable, when the project is combined with government subsidies.

D. Guidelines for Future Inclusive Housing

The section provides guidelines on reducing perceived risk in inclusive housing developments. This maximises the possibility of stimulating funds into residential developments thereby reducing housing shortage in African cities. These guidelines provide fundamental basis for improving future inclusive cities and overall residential developments making housing a viable and sustainable asset. Listed next are the guidelines:

i. The Guidelines

1. Reposition inclusive housing as an investment asset to speed up the supply of housing as it reduces risks of loss. Policies that take inclusive housing as a social-spatial change programme fall short in expediting the supply of the much needed housing in Africa.
2. Collaboration through the 4Ps model is required to deliver inclusive and affordable housing as governments alone have proved to be inefficient, slow, bureaucratic, and often corrupt in providing quality housing that are investment assets. Similarly, the private sector has proven to be too self-serving when contracted for the provision of large-scale housing projects. However, leveraging combined efforts from public, private, and people have the power to build housing developments better than those constructed by government. Through innovative technology, community based approaches such as crowd-funding can be used to create savings towards cooperative housing in Africa.

3. For the new and future inclusive housing in Africa, there is a need to build inclusive housing in Greenfield locations to make them self-sustainable developments. Greenfield location provides a chance to also compensate for the outskirt location by bringing in the preferred amenities. This location makes inclusive housing viable when efficient transport networks link them closer to economic hubs as advocated by Calthorpe (1993). The argument is that complementing inclusive housing with improved infrastructure, transport linkages, business incentives, and supporting social amenities such as clinics, aschools, and parks in the receiving neighbourhoods makes any a location a preferred location that is able to attract higher investment values. This needs to be well designed by planning experts to make sure all components of the inclusive housing are functioning well.
4. To address the social and economic components of inclusive housing, government and municipalities must provide both land and bulk infrastructure, including water and sewer systems to reduce the high cost of these essential services to developers. When government provides these bulk services, the cost of essential services to developers is minimised. This stimulates the private sector to build better housing that maintains investment value in the future and yields competitive financial returns, making house affordable if the value created is shared proportionately. The improved infrastructure, transport linkages, and social amenities do complement the recommended Greenfield location of future inclusive housing through 'land value capture' or planning gain. This location has the potential to circumvent the shortage of land trajectories closer to CBDs in Africa. The argument is that when an inclusive housing development brings its own set of amenities to any location, that location becomes preferable and is capable of attracting higher property values in the future. This makes real estate investors indifferent to where the inclusive housing is located.
5. Housing cooperatives have the power to pool resources together for housing. This can be done by pooling in pension funds, crowd-funding initiatives, and sweat equity as a source of labour helps cut labour costs. This makes the cost of borrowing enhancing to housing affordability. Also, involving people from communities creates entitlement, giving the sense of pride. There is need for comprehensive feasibility studies that reveal preferences for the earmarked inclusive housing developments to be successful. Compensation for house price reduction due to such inclusive housing within the receiving neighbourhoods might not work well for developing countries, due to competing developmental goals that they face. What could work is compensation in the form of improving infrastructure, such as roads to reduce traffic, building new schools, hospitals, and shopping centres to reduce pressure for services and amenities in African cities.
6. Incentive based inclusive housing policy enables a more private sector oriented solution to funding inclusive housing. This requires regulation and monitoring to ensure that the government funding support (contributed value) is converted into social housing units. This means there is need to accurately capitalise such subsidies invested by government into affordable units. If this 'sharing' is not done properly, the created value of inclusive housing investments skews to the private sector, creating a housing shortage conundrum.
7. Improving security services through new efficient police services and neighbourhood surveillance cameras may help curb fears and apprehensions towards inclusive housing, which are mostly occupied by low income people and the associated crime problems.

8. Municipalities must work with asset and property management companies or employ the strategies to access land and bulk infrastructure. Grants such as the Urban Settlements Development Grant (USDG), the Social Housing Regulatory Authority (SHRA) Restructuring Capital Grant, Restructuring Capital Grant Quantum, and Community Residential Units (CRU) can be used to fund inclusive housing.
9. A more income based approach is recommended for African cities, as criterion for developing multi-tenured inclusive housing that do not default into racial demarcations. Intentional mixing of different people may be successful in the 'rented component of inclusive housing' in Africa where young people can learn to work and live together in equal racial inclusive housing environments.
10. In African cities, inclusive housing offers an opportunity for skills development through Small Medium and Micro Enterprises (SMMEs), training young people as building officers, plumbers, bricklayers, and even in the soft skills. This kind of engagement may open employment to help tenants pay rents and mortgage payments to reduce defaulting on payment within the mixed tenure developments. Government should be able to diffuse mortgage default risk by extending guarantee-ship.

ii. Implications for Housing Policy

1. The incentive based financing requires tight monitoring and evaluation to ensure government grants and subsidy-calculated funds get converted into low-income housing within the inclusive housing development. This boosts the much needed supply of housing.
2. There is need for a cautious continuity of future inclusive housing as there still remains individual people who do not believe in mixing poor and rich people, a view held by aristocrats from as early as the 1500s.
3. In the short term, urban regeneration programmes can include use of undeveloped restoration land to build required housing closer to economic hubs, following which Greenfields can be used.

iii. Implications for Housing Market

1. Inclusive developments in Greenfield locations enable building better quality housing designs which are able to maintain value in the future. This reduces risk of loss when house owners default due to loss of value.
2. Competent asset and property management companies can help municipalities collect more revenue from the 'new smart inclusive' cities.

E. Discussion Questions

1. Are there notable improvements on people living in current inclusive developments and cities in Africa?
2. How have the 'Tax Increment Financing' (TIF) and 'Equity Tax' been successful in building the required infrastructure in Africa?

3. How can we increase availability of land for inclusive, affordable, and smart housing in African cities?
4. Is there a fool-proof plan by housing development banks in Africa to reduce mortgage default risk by low income households?
5. How best can we model the development of future inclusive housing in an African city?

References

Becker, D. A (2020). The Meaning of Belonging: Race and the Making of South Africa *Neoliberalism and the State of Belonging in South Africa*, 81–123.

Bridge, G., Butler, T., & Lees, L. (2015). Mixed Communities: Gentrification by Stealth? *The Journal of Urban Affairs*. The Policy Press, 2012, 37(5), 658–659.

Calthorpe, P. (1993). *The Next American Metropolis, Ecology, Community and the American Dream*. New York: Princeton Architectural Press.

Charlton, S. (2004). *An overview of the Housing Policy and Debates, Particularly in Relation to Women (or Vulnerable Groupings)*. Johannesburg, South Africa: Centre for the Study of Violence and Reconciliation. pp. 1–38.

Chegut, A., Eichholtz, P., & Holtermans, R. (2016). Energy Efficiency, and Economic Value in Affordable Housing. *Energy Policy*, 97, 39–49.

Chigwenya, A. (2019). Financing Low-income Housing in Bulawayo, Zimbabwe: Implications for the Right to the City and Inclusivity. *Urban Development Issues*, 1(64), 39–48.

Department for International Development (DFID) (2010). Government of UK Evaluation of Community Led Infrastructure Finance Facility. (CLIFF) India Country Report. Research for Development Output.

Available at: Evaluation-of-Community-Led-Infrastructure-Finance-Facility-CLIFF-India-country-report—2010.aspx. [Accessed: 05 August 2017].

Eziyi, O. I. (2012). *Public-Private Partnerships (PPPs) in Housing Provision in Ogun State, Nigeria: Opportunities and Challenges*. Canaan Land, Ota, Ogun State, Department of Architecture, College of Science, and Technology Covenant University. Available at: https://www.academia.edu/7077154/Public_Private_Partnership_PPP_in_Housing_Provision_in_Lagos_Megacity_Region_Nigeria. [Accessed: 17 May 2019].

Ferilli, G., Sacco, P. L., Tavano Blessi, G., & Forbici, S. (2017). Power to the People: When Culture Works as a Social Catalyst in Urban Regeneration Processes (And When It Does Not). *European Planning Studies*, 25(2), 241–258.

Graham, N. (2016). Financing Infrastructure for Housing Developments: Case Studies from Sub-Saharan Africa. Center for Affordable Housing (CAHF) Johannesburg. *Case Study Series*, 4, 2–22.

Gruis, V., & Nieboer, N. (2004). Strategic Housing Management: An Asset Management Model for Social Property Owners. *Property Management*, 22(3), 201–213.

Hulse, K., & Yates, J. (2017). A Private Rental Sector Paradox: Unpacking the Effects of Urban Restructuring on Housing Market Dynamics. *Housing Studies*, 32(3), 253–270.

Jones, G. A., & Datta, K. (2010). Enabling Markets to Work? Housing Policy in the 'New' South Africa. *International Planning Studies*, 5(3), 393–416.

Jupp, B., Sainsbury, J., & Akers-Douglas, O. (1999). *Living Together: Community Life on Mixed Topology Estates*. London: Demos.

Khadduri, J., & Martin, M. (1997). Mixed-Income Housing in the HUD Multifamily Stock. *Cityscape: A Journal of Policy Development and Research*, 3(2).

King, R., Orloff, M., Virsilas, T., & Pande, T. (2017). *Confronting the Urban Housing Crisis in the Global South: Adequate, Secure, and Affordable Housing Confronting the Urban Housing Crisis in the Global South*. Available at: www.citiesforall.org.

Massyn, M. W., McGaffin, R., Viruly, F., & Hopkins, N. (2015). The Challenge of Developing Higher Density, Affordable Housing in the Inner City of Cape Town. *International Journal of Housing Markets and Analysis*, 8(3), 412–428. https://doi.org/10.1108/IJHMA-11-2014-0049.

McGreev, M. P (2018). Housing Diversity and Affordability: The Effects of 35 Years of Exclusionary Land Use Regulations on Housing Affordability in Adelaide, South Australia. *Urban Policy and Research*, 36(3), 336–353.

Miller, J. (2016). Inequality in Africa. Unequal Scenes Neighbourhoods. Available at: https://www.behance.net/gallery/38385837/Kya-Sands-Bloubosrand. [Accessed 15 May 2018].

Mohamad, Z. Z., Yang, F. C., Ramendran, C. S. P. R., Rehman, M., Nee, A. Y. H., & Yin, Y. C (2020). Embedding Eco-Friendly and Smart Technology Features in Affordable Housing for Community Happiness in Malaysia. *GeoJou rnal*. https://doi.org/10.1007/s10708-020-10247-8.

Montgomery, N., Squires, G., & Syed, I. (2018). Disruptive Potential of Real Estate Crowdfunding in the Real Estate Project Finance Industry: A Literature Review. *Property Management*, 36(5), 597 –619. https://doi.org/10.1108/PM-04-2018-0032.

Moss, V., Dincer, H., & Hacioglu, U. (2013). The Nature of the Creditor-Debtor Relationship in South Africa. *International Journal of Research in Business and Social Science*, 2(2), 2147–4478.

Myeni, S. L., & Mvuyana, B. Y. C. (2018). Participatory Processes in Planning for Self-Help Housing Provision in South Africa: Policies and Challenges. *International Journal of Public Policy and Administration Research*, 5(1), 24–36.

National Department of Human Settlements (2016). *National Department of Human Settlements Strategic Plan: 2015–2020*. Pretoria, pp. 1–44. Available at: http://pmg-assets.s3-website-eu-west 1.amazonaws.com/DHS_FINAL_REVISED_STRATEGIC_PLAN_2015_2020.pdf. [Accessed 06 May 2017].

Onatu, G. O. (2012). Sustainable Land Use and Development: Perspective on Cosmo City, Johannesburg, South Africa. *Sustainable Development and Environmental Protection*, 2(1), 67–77.

Osman, A., Sebake, N., & Davey, C. (2011). Focus on Medium-Density Mixed Housing: An Important Component in the Transformation of South African Housing Environments. 3rd CSIR Biennial Conference 31 August & 01 September 2010 CSIR International Convention Centre Pretoria, pp. 1–22. Available at: https://researchspace.csir.co.za/dspace/handle/10204/2516. [Accessed: 20 September 2017].

Simbanegavi, P. (2021). Year Book, South Africa, Center for Affordable Housing Finance (CAHF). South Africa Country Profile. Available at: https://housingfinanceafrica.org/resources/yearbook/

Simbanegavi, P., Shani, Z., Watkins, J., & Ramruthan, K. (2019). Making Rental Housing in the Gap-Market More Affordable through Green Building Technology. In C. Aigbavboa & W. Thwala (eds.), *The Construction Industry in the Fourth Industrial Revolution*. Cham: CIDB and Springer, pp. 241–251.

South African Property Owners Association (SAPOA) (2018). *Inclusionary Housing, Towards a New Vision in the City of Jo'burg and Cape Town Metropolitan Municipalities*. Johannesburg: SAPOA, pp. 4–47.

Available at: http://www.sapoa.org.za/media/2948/inclusionary-housing_revised.pdf. [Accessed: 03 December 2018].

Taruvinga, B. G., & Mooya, M. M. (2018). Neo-Liberalism in Low-Income Housing Policy-Problem or Panacea? *Development Southern Africa*, 35(1), 126–140.

Thompson, L. A. (2001). *A History of South Africa*. 3rd ed. Boston: Yale University Press.

Tunstall, R., & Fenton, A. (2006). *In the Mix, a Review of Mixed Income, Mixed Topology, and Mixed Communities: What Do We Know*. New York: Housing Corporation, Joseph Rowntree Foundation, & English Partnerships.

UN-HABITAT (2014). *Stocktaking of the Housing Sector in Sub-Saharan Africa Summary Report*. Washington, DC: The International Bank for Reconstruction and Development. Available at: http://www.worldbank.org/content/dam/Worldbank/document/Africa/Report/stocktaking-of-the-housing-sector-in-sub-saharan-africa-summary-report.pdf. [Accessed: 07 June 2017].

United Nations (2019). Sustainable Development Goals (SDG Compass). Available at: https://sdgcompass.org/sdgs/sdg-11/. [Accessed: 11 July 2019].

Van Gent, W. P. C., & Musterd, S. (2013). Unintended Effects of Urban and Housing Policies on Integration: "White" Discontent in the Dutch City. *Geography Research Forum*, 33, 64–90.

World Bank (2018). *The World's Largest Slums: Dharavi, Kibera, Khayelitsha and Neza, Nairobi*, pp. 1–7. Available at: https://www.habitatforhumanity.org.uk/blog/2017/12/the-worlds-largest-slums-dharavi-kibera-khayelitsha-neza/. [Accessed 19 March 2017].

11

A TALE OF TWO PATHWAYS

How the low-end Gauteng residential property market may be shaped either by state-driven supply or by a more enabled private sector

10 December 2020 [revision 21 January 2021]

Napier Mark, Cooper Antony K, Holloway Jennifer P, Thanjekwayo Londeka, Graham, Nick and le Roux Alize

1. Background

> Housing projects are not reducing the percentage of households in informal dwellings. The [South African General Household] Survey shows that 81.1% of all households resided in formal dwellings in 2018. Although the percentage of households that have received some kind of government subsidy to access housing has increased from 5.6% in 2002 to 13.6% by 2018, 13.1% of households were still living in informal dwellings. This could be attributed to the fact that rapid household growth and population relocation is making it very difficult to address existing backlogs in the face of fresh demands.
>
> (StatsSA 2019)

1.1. Introduction

The chapter will discuss some of the ways in which the South African government may be able to use the tools at their disposal to encourage greater private sector engagement and activity. This chapter discusses the findings on housing demand and supply, from the *Gauteng Housing Demand Model* project, conducted for the Gauteng Department of Human Settlements (GDHS) (Napier *et al* 2016). It also presents alternatives for housing delivery, depending on policy and budgetary priorities. The model can test these alternatives and the likely impacts of policy decisions now and into the future. The initial models were run in 2016, using a 'medium' forecast for population and economic growth, as discussed in Section 2. Subsequently, of course, the countrywide, extended lockdown due to the pandemic of SARS-CoV-2 (severe acute respiratory syndrome coronavirus 2) and the disease it causes, COVID-19 (coronavirus disease 2019) (Paules *et al* 2020), has disrupted lives and livelihoods, and altered the economic outlook considerably. The possible ramification of this unfolding economic outlook has been considered and the likely housing implications resulting from the revised growth scenarios are discussed in Section 4.

DOI: 10.1201/9780429279256-11

1.2. Gauteng province

Gauteng is one of nine provinces in South Africa (Figure 11.1) and includes three of the eight metropolitan municipalities (metros), namely Johannesburg, Tshwane and Ekurhuleni, and two District Municipalities, Sedibeng and West Rand. While the smallest province in area, it accounts for one third of the national gross domestic product (GDP) (StatsSA 2020c) and 26% of the national population: according to the *Mid-year Population Estimates* for 2020, there are about 59.62 million people in South Africa, with about 15.5 million in Gauteng (StatsSA 2020b). As at June 2021, 4.648 million people were employed in Gauteng, being 31% of the total employment in South Africa (14.942 million) (StatsSA 2021). Gauteng has the lowest poverty level of the nine provinces – yet it is as high as 33%. Despite housing a large share of the country's population, the province has the lowest average people per dwelling size at 3.01 persons per dwelling, compared to the 3.45 national average (StatsSA 2019).

While Gauteng still has significant agricultural areas, Gauteng is mainly an urban province and is often referred to as the Gauteng City-Region (GCR). The GCR is also considered to extend into the neighbouring local municipalities (Mabin 2013), but as they lie outside the jurisdiction of the Gauteng government, they have not been included in the analysis reported herein.

1.3. Housing

With very few traditional dwellings compared to other provinces, Gauteng nevertheless has the highest percentage of informal dwellings (including temporary backyard structures, or 'shacks') at 19.8%, compared to the national average of 13.1% (StatsSA 2019). Government housing policy and programmes view inadequate housing, as defined in the South African Constitution, (South Africa 2012) as part of the backlog that the state needs to address in its social housing programmes (Tissington 2010). As stated in the Housing Act regarding the responsibilities of the State:

> The establishment and maintenance of habitable, stable and sustainable public and private residential environments to ensure viable households and communities in areas allowing convenient access to economic opportunities, and to health, educational and social amenities in which all citizens and permanent residents of the Republic will, on a progressive basis, have access to: permanent residential structures with secure tenure, ensuring internal and external privacy and providing adequate protection against the elements; and potable water, adequate sanitary facilities and domestic energy supply.
>
> *(South Africa 1997)*

The number of households living in informal housing located in informal settlements plus the number of households living in overcrowded conditions (e.g. in backyard dwellings) are often used as a proxy for the extent of housing need in any province (and city or town) at any given time.

Gauteng also has by far the highest percentage of backyard dwellings (24%), compared to the national average of 15% (StatsSA 2019). The other important measure to understanding housing demand is whether people rent or own their dwellings. Whether renting structures in backyards, or renting houses or apartments, the proportion of rental housing stock is the highest when compared to other provinces. This proportion of rented housing stock,

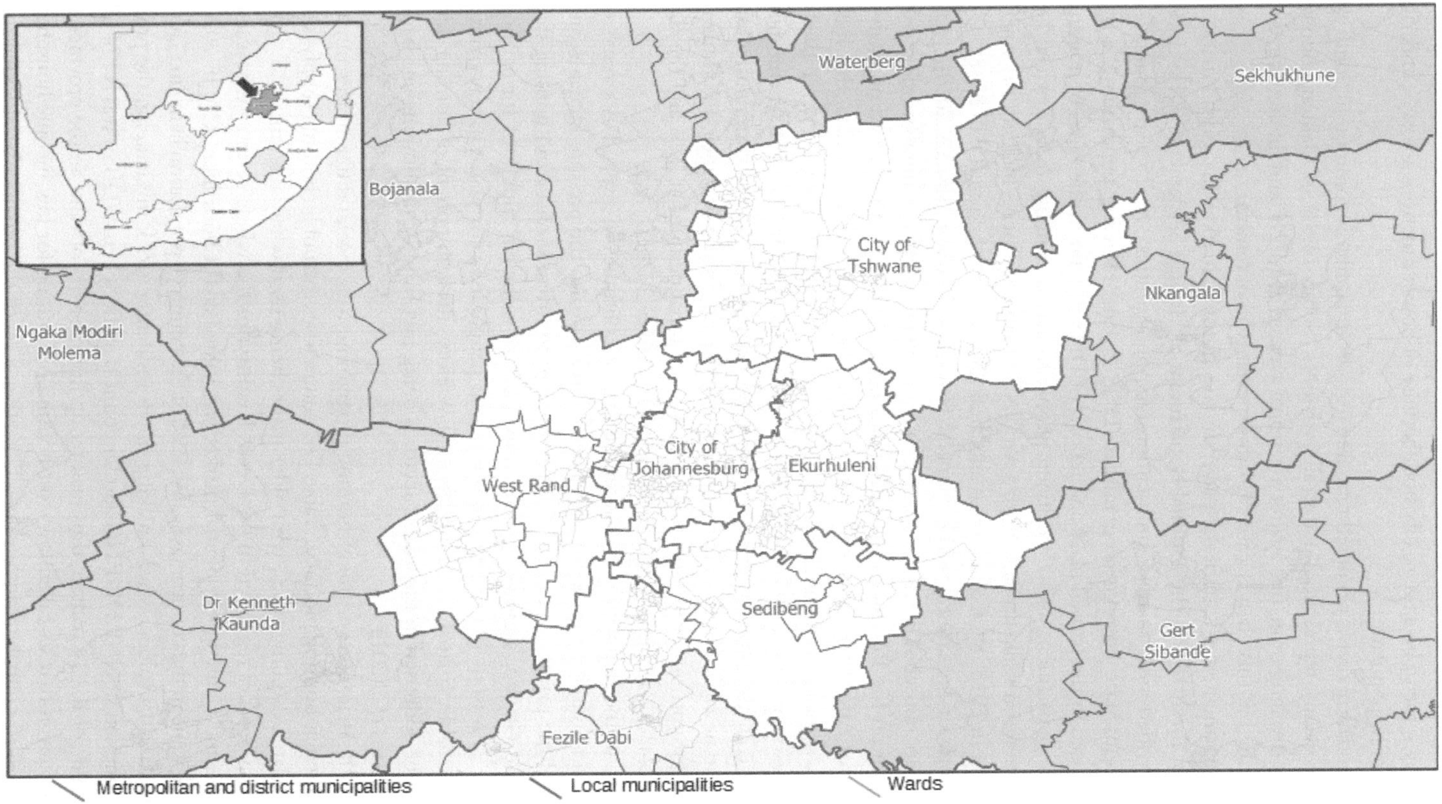

FIGURE 11.1 Gauteng municipalities and wards (data obtained from MDB (2016))

whether formal or informal, is 37% against the national average of 26.5%. Again, this makes sense for a rapidly growing province that, even though covered by five municipal areas, is considered by some to be a mega-city both because of how it functions across boundaries and for the size of its population.

Accommodation that is occupied rent-free (e.g., from family) is common in South Africa, and is fairly evenly represented across all provinces. Some 14.8% of Gauteng households occupy their dwellings rent-free against a national average of 14.2%, and the housing demand model described in this chapter caters for this special type of tenure.

The government is a significant player in the housing market in South Africa. In its annual report for 2018/19, the national Department of Human Settlements (DHS) indicated they had provided more than 4.7 million housing opportunities over 25 years country-wide, through various state housing programmes (DHS 2019). 'Housing opportunity' is a term commonly used in the formal plans and reports of DHS to refer to the types of housing and land that the department assists qualifying people to access, ranging from serviced land to rental apartments in walk-up flats. With a national human settlements budget (for projects) of R18.7 billion in 2018/19, Gauteng was allocated 28% or R5.1 billion (DHS 2018).

At the time of the latest available General Household Survey (StatsSA 2019), 13.6% of South African households self-reported they were living in government-subsidised housing. Focusing on the Province of Gauteng and using another method based on analysis by the Centre for Affordable Housing Finance of the 2019 Deeds Registry data, that included subsidised Reconstruction and Development Programme (RDP) (South Africa 1994) and Breaking New Ground (BNG) (DHS 2004) houses, and discount-benefit scheme housing (see Table 11.2), the proportion of residential properties that were government subsidised was estimated at 27%. For the metropolitan municipalities in the province, Ekurhuleni had the highest proportion at 28% of stock, followed by Johannesburg at 27% and the City of Tshwane at 24% (Namponya & Nkhonjera 2020a, 2020b, 2020c).

However, as the population of Gauteng Province continues to expand as a result of in-migration and household growth (Peberdy *et al* 2017), so also the pressure increases on government and the private sector to enhance and expand urban services and housing stock to accommodate Gauteng's growing population. Lack of sufficient supply of formal housing for a variety of reasons (Rust 2006) means that as people look for places to stay many cannot find affordable, formal accommodation. Therefore, the component of housing demand that cannot be addressed in the formal sector is being met in informal housing and backyard dwellings (Mans *et al* 2014).

The South African government housing programme uses several different instruments to produce different housing products. Table 11.1 summarises the public housing supply, subsidy types and related dwelling types, with older names being used where relevant.

In 2017, properties registered in the Deeds Registry for the whole country were valued at R5.1 trillion and Gauteng's residential market was valued at R1.96 trillion (CAHF 2020), or 38.5% of the value in the country (as against 26% of the country's population living in Gauteng).

Moving to 2019 values, when breaking down residential stock in Gauteng, the property value of subsidised stock (RDP, BNG and Discount Benefit Scheme (DBS)) was R174 billion of a total value of R2.14 trillion, thus making up just over 8% of value in the province. This is significant given that so-called low-cost housing or housing for low income households is often assumed to contribute little value. The other 92% of the province's residential market

TABLE 11.1 *Public housing supply, subsidy types and related dwelling types*

Demand-side Programmes *Facilitates access to products available on the market*[1]	*Incremental Housing Programmes* *Facilitates access to 'housing opportunities'*[2] *through a phased process*	*Social and Rental Housing Programmes* *Facilitates access to rental housing opportunities to support urban restructuring and integration*
– Finance Linked Individual Subsidy Programme (FLISP)[3] – Individual Subsidy (Non-credit linked)[5]	– Integrated Residential Development Programme (IRDP) – Enhanced Peoples' Housing Process (PHP) – Informal Settlement Upgrade Programme (ISUP) – Project Linked Consolidation Subsidies Programme (CSP) – Rural Subsidy Programme	– Social Housing Programme (SHP) – Institutional Subsidy – Community Residential Units Programme (CRUP) – Backyarder programmes (local, *ad hoc*)
Dwelling types typically produced		
• Bonded, partly subsidised individual or semi-detached houses for ownership	• 'RDP' or 'BNG' houses[4] on plots for ownership • New build flats for ownership • In situ upgraded informal settlement with original informal dwellings, or in later phases with 'top structure'. • Serviced sites (Greenfields) for ownership • Owner-built houses on serviced sites (e.g. ePHP)	• New build flats for rental • Upgraded Council rental walk-up unit • Upgraded hostel unit • Backyard, rented room, serviced

Notes:

1 Or as the Housing Code 2009 puts it, "This Programme intends to give effect to the policy objective to stimulate the secondary housing market" (DHS 2009).

2 'Housing opportunities' is a term commonly used in the formal plans and reports of the Department of Human Settlement that refers to the types of housing and land that they as a department assist qualifying people to access, from serviced land to rental apartments in walk up flats.

3 State-provided subsidy used to pay the deposit on a house or to decrease the size of the home loan (Mathibela 2019).

4 Housing provision supported by the state was a significant component of the Reconstruction and Development Plan (RDP) (South Africa 1994). The free-standing houses on land parcels provided free for ownership, which was the most prolific output of the first 20 years of the housing programme, came to be known in common parlance as RDP houses. From 2004 onwards, following the publication of the Breaking New Ground (BNG) plan by DHS (2004), and a larger subsidy for a housing product built to comply with new National Building Regulations, the houses produced from then came to be known as 'BNG houses' (Landman & Napier 2010).

5 Non-Credit Linked individual subsidies are available to persons meeting the qualification criteria and who do not qualify for credit from a financial institution. The purchase of an existing house is therefore made solely from the subsidy amount awarded (Part 3 of the National Housing Code (DHS 2009)).

Note that this table excludes some subsidies that are no longer in common use, along with the Emergency Housing Assistance Programme as it is a funding mechanism that only arises in the event of an emergency or disaster and delivery targets are not set in place ahead of time.

in 2019 was therefore un-subsidised and contributed by the commercial sector. The three metropolitan municipalities speak for 91% of the residential property value in the province, with the two, more sparsely populated, district municipalities accounting for the balance.

Coupled with the housing background sketched here, Gauteng Province is also expected to continue to experience rapid urbanisation and population growth in the next three decades. Current population forecasts project that between 20.30 and 21.79 million people will live in the province by 2050 (le Roux *et al* 2019). It is in this vibrant, growing economic and social context that the discussion of the future of housing demand and supply is particularly pertinent, and the policy and programme decisions made today will have long-lasting implications for the residents of Gauteng and an effect on the broader regional economy.

1.4. Models

By way of introduction, in a very general sense a model is a representation of a real-world object or process captured with the intention of simulating real-world behaviour and responses to stimuli in a controllable environment. More specifically, in an economic or mathematical context it refers to one or more algorithms, typically compiled into a software tool, which can provide quantitative outcomes for the purposes of decision making. Time-based models have the benefit of being able to project into the future by using information gathered from past behaviour and can be used for projecting various outcomes based on both static input data and scenario parameters that are adjusted by the user. Models used for the purpose outlined in this chapter are very beneficial to planners as they can provide several possible futures, depending on different foreseeable behaviours or interventions implemented. The decision maker does therefore not have to blindly adopt a 'wait and see what happens' attitude but can be better equipped to know the expected outcome of taking certain actions.

'Demand' models strive to predict the future demand for goods or services by looking at influencing factors and determining consumer choice behaviours from past data. Many housings demand models have been created in an attempt to capture consumer preferences to understand housing consumption (see for example: Boumeester (2011), Zabel (2004), Suárez and Becerra (2015), with many of these focused on the effects of house price on demand (see for example: Pain & Westaway, 1997; Bajari *et al* 2013). Another set of models which are fundamental to the study of economics are 'supply and demand' models. Traditionally, these models focus on finding a suitable price that will provide an 'equilibrium' between the goods being supplied and the quantity demanded (see for example Wang and Zhang (2017)). However, these models are typically used for a single product where that product provides for a basic want rather than an essential need, such as housing. In a real estate context, this type of model applies at different price levels of housing demand for the various income levels, and therefore supply needs to match affordability at these different levels to reach equilibrium. The Gauteng Housing Demand Model (Napier *et al* 2016) used in this chapter draws on aspects from both these types of models by predicting housing demand into the future, based on: affordability, housing preferences and the basic need for shelter whilst also looking at which scenarios of future supply can help achieve equilibrium.

This chapter describes the design of the demand model based on local and international theory, which emulates housing-choice behaviour based on past trends and building on a number of key assumptions. After outlining the model methodology, this chapter draws on the data, forecasts and scenarios generated by the Gauteng Housing Demand Modelling

Project (Napier *et al* 2016), to discuss possible alternative futures. This chapter closes with a discussion about the tools and approaches that especially state actors might consider in mobilising greater private sector engagement at the lower end of the residential market.

2. Gauteng Housing Demand Model

The Gauteng Housing Demand Model (Napier *et al* 2016) has as its objective the description of housing supply, demand and backlogs in the municipalities that make up the province. Figure 11.2 shows the dashboard or front-end for using the model, which allows key input parameters to be altered, particularly the demand and supply parameters, without changing the baseline data. However, the baseline data can still be altered, if required. The dashboard makes key outputs visible, showing how they change according to the changes to the input parameters. Hence, it allows input scenarios to be tested, be they high, medium or low.

The situation that the model describes reflects a baseline reality (observed reality) derived from census numbers, real residential structures that have been built and the values of properties according to municipal valuation rolls. The census data about population demographics, household characteristics, dwelling types and income (amongst other data variables) were sourced mainly from Stats SA products and other tools (such as the Quantec EasyData Service (Quantec 2020)) that enhance the Stats SA data. The historical data that were used to populate the model draw from a combination of the 1996, 2001 and 2011 censuses extracted from the SuperCROSS database (StatsSA 2020) and the geographically re-aligned data from the EasyData Service (Quantec 2020), so that it is possible to identify longer term trends over a 15-year period leading up to the present.

The model then projects forwards from 2011 to the year 2030 with the objective of understanding the likely future trends in the residential sector. Here the model uses a number of assumptions about the future and draws on an algorithm to allocate housing stock to households in the future.

The forecasts to 2030 are projections of possible futures; however due to the uncertainty associated with long-term projections, there are many factors that need to be considered that could affect housing demand and supply in the future. To accommodate these uncertainties, the model was built to allow the user to manipulate several influential factors to see the effect of various external influences on the housing sector, and to model the effect of different decisions being made by the Gauteng Provincial government and other actors in the public and private sectors. The main external factors that will affect future projections are:

- demographic variables in the province, for example:
    - population growth projections;
    - household size projections;
    - household income bands;
- economic growth projections.

To illustrate this, the model currently uses default figures for population that were independently generated (IHS Markit 2012) and that show a 2030 Gauteng population of 19.4 million people by 2030. The model is designed to reflect what could happen if there were high, medium or low population growth in Gauteng. The default (or 'medium') forecast that is

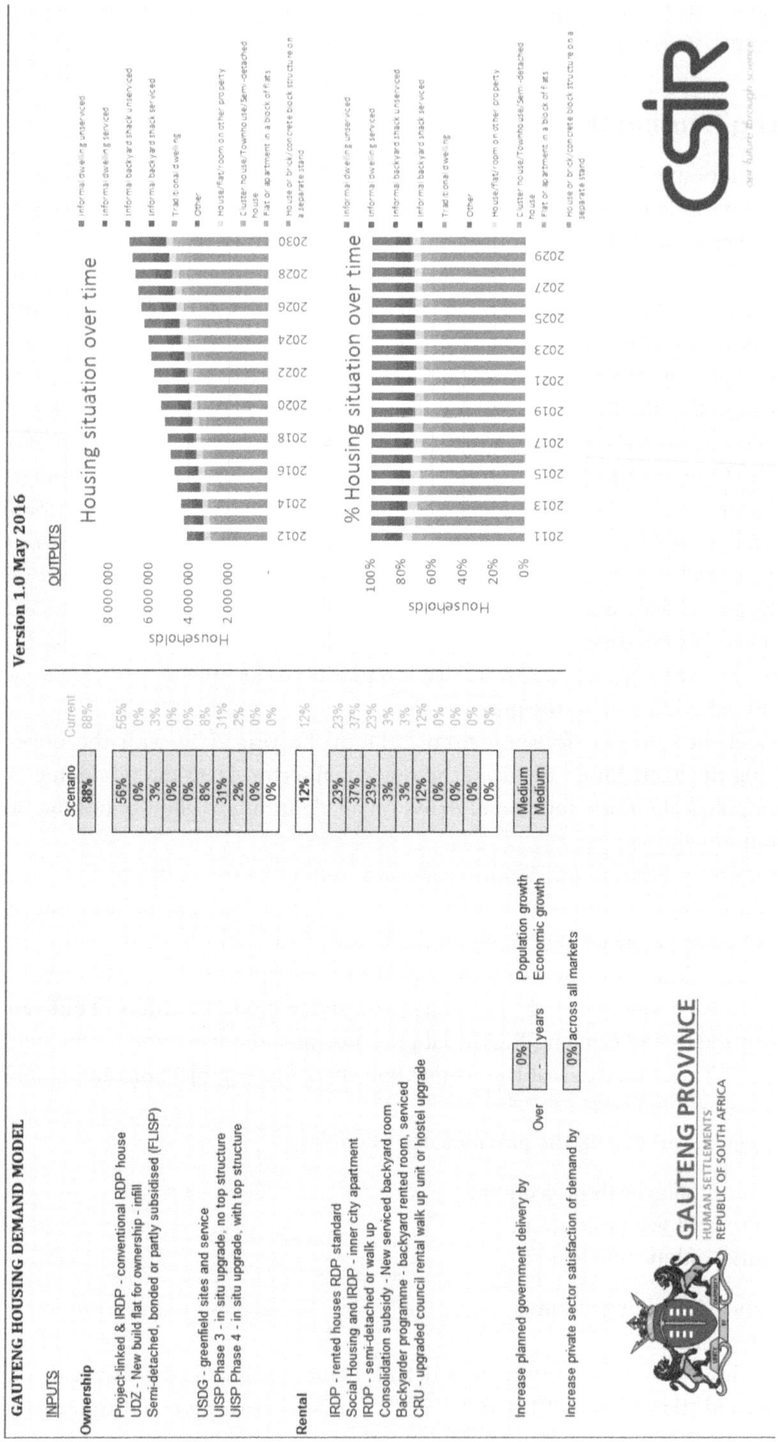

FIGURE 11.2 Gauteng Housing Demand Model dashboard

used to inform housing demand is based on a projection of 19.4 million people by 2030. The high population projection would adjust this upwards by 1% per annum and the low projection would adjust it downwards by 1% per annum, both from a baseline of 2012.

The model also allows the user to model high, medium or low economic growth into the future, which affects incomes and affordability on the demand side, and the rates of housing production on the supply side. The baseline model uses the medium projection forecast as the default situation. All three spheres of government have some control (direct or indirect) through policies and decisions on the main internal factors, such as:

- the size of state housing and infrastructure budgets;
- the scale of housing supply, including a range of types of housing opportunities, and tenure;
- options (ownership or rental);
- the mix of types of housing opportunities directly delivered or incentivised by government;
- the cost per housing opportunity in the various categories of state supported supply;
- the density of settlements built, and associated land required; and
- private sector supply, through incentives and demand side instruments such as FLISP (Finance Linked Individual Subsidy Programme).

These internal and external factors (demographic and economic) that affect future forecasts can all be manipulated in the model. The most important input parameters can be altered directly in the front-facing dashboard. Each metro and DM has its own model, but there is also a combined Gauteng model, which is the one utilised for this chapter. The model can also be calibrated through populating additional input parameters as they become available, such as future censuses and mid-year population estimates; inflation and growth rates; building costs and property values; and updated projections. This will improve the accuracy of the model as the situation unfolds going forward from the present, as discussed in Section 4 for dealing with the impacts of the COVID-19 pandemic.

The next section will unpack the results from running the model with the following three scenarios:

- the business-as-usual scenario, where both the state and the private sector continue to follow their current housing delivery trends, using the medium population and economic growth forecasts, as used in 2016;
- the increased private sector scenario where the private sector plays a more significant role in providing affordable housing, using the medium population and economic growth forecasts, as used in 2016; and
- the impact of the COVID-19 lockdown during 2020, still using a medium population forecast but with a low economic forecast.

It should be noted that the model itself is essentially non-spatial, other than describing the current and future situations for each district and metro municipality. Many of the inputs (e.g. where people live, or property values in different parts of Gauteng) can be mapped, but the model itself cannot predict future spatial trends as a direct output of the model. For example, the geographical location of future state-funded delivery is not integrated directly into the model. This would need to be a separate mapping exercise, employing different algorithms to predict the finer-scale locational decisions. However, the state land needs (hectares) and housing delivery costs in different municipalities can be extrapolated from the model as it currently stands. Different housing types have different land needs because of varying built densities.

The model provides for several dimensions for backlogs that may be measured in the present and into the future, that is, the quantity of housing needed for households currently in inadequate housing, be they informal, over-crowded, in poor condition, etc:

- **The number of households or individuals qualifying to make claims on the state for housing assistance**: this is modelled through the user being able to choose the income groups targeted by the various housing policies and instruments.
- **Owned vs rental accommodation**: if the province scales up the production of fully or partially state-funded rental accommodation (as opposed to houses for ownership), then by modelling who qualifies for that stock, the costs of production and the proportion of state funding that would go to that instrument, it is then possible to understand the impact of that policy decision on future backlogs which would manifest in the number of people applying for subsidies or other instruments, as reflected in a housing demand database.
- **The backlog of adequate housing units**: this is essentially about affordability of housing units in the different categories of housing type (and related property values). In the model this is reflected as the gap between housing demand (effective demand in this case), and current and future housing supply in the province. An adequate housing unit (for rent or ownership) costs a certain amount to produce. Working up from a basic unit (or indeed a serviced erf (plot) without a house), there is then a range of stock available according to what households can afford and choose to pay for. If it is projected that the formal system will not supply the range of adequate land and housing opportunities to cope with social and effective demand, then the model allocates future households to inadequate housing situations (simplified into shacks in informal settlements and backyard dwellings and overcrowded formal stock). Then in terms of what is defined in policy as inadequate housing situations, this becomes the size of the backlog into the future and is a very important indicator of whether the public and private sectors are acting in ways that are likely to keep pace with population growth in the province given various economic growth outlooks.
- **Inadequate housing units**: in policy terms, the number of inadequate housing units (that is, those that are poorly built) may be seen as a secondary definition of backlog. However, it is important for the whole housing sector because whether a household living in inadequate housing qualifies for government assistance (i.e., to make a claim on the state) or not, the Constitution obliges society to address inadequate housing. In Constitution and policy terms, this is probably the primary definition of backlog for the government.
- **The land backlog and land 'need'**: the backlog projections described earlier are also accompanied both by an estimate of the cost to the state of addressing social demand for adequate housing, and an estimate of the amount of land needed to build what the government is planning to build or initiate. The overall land need is projected according to the chosen mix of housing typologies, which are built at different densities. To apply

this more directly, it would be advisable to take the overall projected land need from the model, and then to deduct the area of land that the province and possibly the municipalities have already acquired and earmarked for targeted low income housing or serviced land release. The difference would indicate what area of land still needs to be acquired, or unlocked through joint ventures or agreements with current private or public landowners. This would then be the land backlog.

- **Delivery backlog**: the delivery backlog is the difference between what the province may have planned to deliver (the delivery targets are entered into the model to calculate future public supply) and what has actually been delivered. This would only become evident later on, as housing opportunities and units delivered are reported on, and measured against the original targets. The term backlog does imply some commitment (on the part of the state, or of society more broadly) to reduce the number of inadequate housing situations, or inadequate water and sanitation levels, or the like. In that sense, backlogs are defined against policy commitments as the current shortfall that needs to be addressed.

With favourable economic growth, incomes and thus affordability would improve, and less people would be dependent on state subsidised housing stock or serviced land. In the case of incremental building processes, where people may benefit from services and then build their own housing over time, this is factored into the model by separating serviced informal settlement units or backyard dwellings, from unserviced units. Once informal land or housing is serviced to a certain level, then the backlog of adequate housing units is partly met, and the incremental building process will complete the move towards adequate housing. The numbers of each category are evident from the model.

Here, we have introduced how the model generates the findings that will be discussed in the rest of this chapter. As mentioned, findings from the medium growth forecast will be discussed along with the findings of the scenario that would emerge if there were to be variations in the demographic and economic forecasts. Scenarios will then be discussed that relate to how the province may decide to shift policy and allocate budget and effort in the future. Not all policy scenarios can be entertained, and the objective is to demonstrate how the model may be used to model different alternatives rather than to comprehensively describe a great number of these alternatives. Finally, we discuss the impact of the COVID-19 lockdown on the scenarios.

Figure 11.3 provides a high-level overview of the model, with the aim of achieving equilibrium between the demand for, and supply of, housing. From the existing housing situation, the household dynamics (such as population growth forecast and changes in household size) produces the uninfluenced demand (how many additional households require housing, before supply is considered). Similarly, the existing housing stock and the supply dynamics create the new (or updated) housing stock. With the housing adjustment, the demand and stock result in the new housing situation and the new equilibrium. This is repeated for each year in the model.

The mechanisms through which households in need of housing are allocated to the modelled available housing stock is termed a housing adjustment 'pseudo decision algorithm', see Figure 11.4. The algorithm is applied stepwise for each household income band, starting with the highest income households, who exercise the most choice, and ending with the lowest income band. Once the highest income band has taken up housing stock, the next income band gets to choose, thus explicitly placing the negative impacts of a shortage of housing

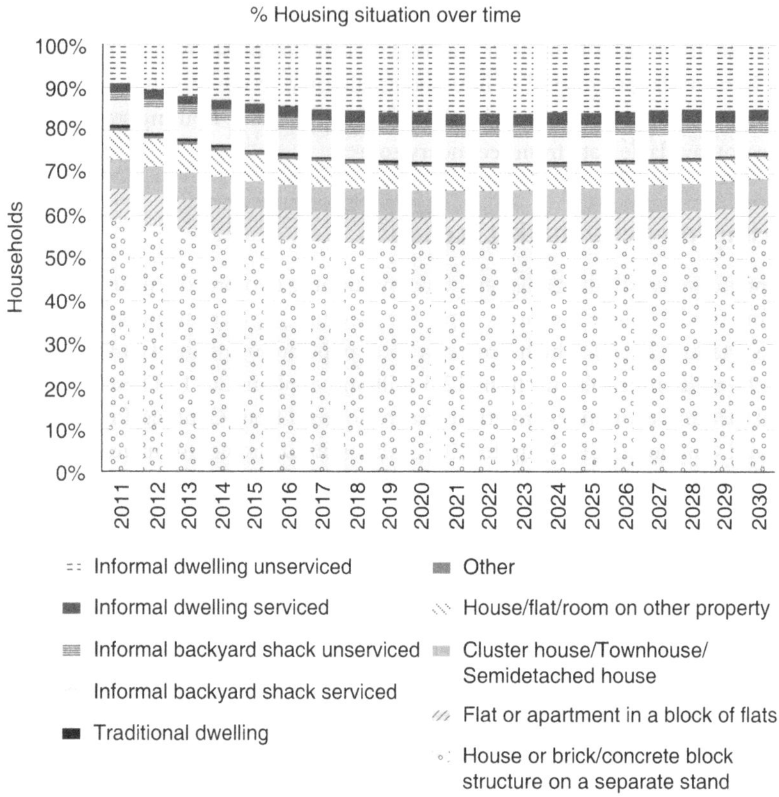

FIGURE 11.6 Overall change in the Gauteng housing situation

Figure 11.5 shows the increasing trend in demand for adequate housing each year (excluding serviced informal dwellings from this demand), reaching levels of over 1.4 million by 2025 and with a marginal decline in demand only starting in 2028. In addition, this scenario shows (see Figure 11.6) that without any significant changes to State housing programmes or the behaviour of private sector housing actors, and assuming medium economic and population growth, the proportion of Gauteng households living in informal housing and backyards will increase from 20% in 2011 to 26% by 2030. In number terms this means an increase from 787,928 households to a total of 1.82 million households living in informal dwellings (some serviced, and many not).

Essentially, this scenario indicates that if the state wishes to continue on this housing delivery pathway and the state aims to reduce or remove the housing backlog in the province (i.e. that all residents live in adequate housing of some kind, without significant levels of overcrowding), then the state budget required to achieve this end would have to be drastically increased. If the state budget or delivery mechanism associated with housing stock is not changed drastically this would result in a much more informal Gauteng province by the year 2030, resulting in the failure of current goals and objectives set by the government.

3.2. Scenario 2 – increased private sector involvement

This second, contrasting scenario tested in the demand-supply model considers what might happen if the state is to step back to some degree, and view the private sector as a meaningful partner with its own strengths, and of course its own inherent limitations. If the state were to work at creating the conditions in which the private sector would be more willing to invest, develop and finance further down-market than it has previously, then this would create a significantly different housing outcome in the province.

Private sector supply was modelled through assuming that the private sector can meet a certain percentage of demand. In the business-as-usual scenario, the private sector supply does not meet all demand (as seen in the so-called gap market, comprising households who cannot afford a house on the market but also do not qualify for government subsidy, see the Financial and Fiscal Commission brief FFC (2012)). For Scenario 2, an assumption was made that the private sector would meet 100% of demand by 2022 in income bands above a monthly income of R3 500.00 (using the 2011-rand value). Historical figures of private sector supply remain up to 2016 in the model after which the private sector starts increasing their response to demand for all income groups from the gap market upwards. The model does not consider how this might be done, or the costs thereof, but a range of policy and other stimuli are discussed in the recommendations.

Figure 11.7 shows how the response by the private sector results in a rapidly decreasing demand to levels of just over 600 000 in 2030. This is achieved in this scenario through the private sector's supply of affordable houses at lower income levels which reduces the backlog of households requiring adequate housing in each forecast year, thus reducing the total demand. Similarly, Figure 11.8 shows how the private sector can arrest the rapid growth of informality, resulting in an overall gradual percentage decrease (across all housing in Gauteng) from 2018 onwards to a value of 15% by 2030, compared to 26% without private sector involvement. The model shows that in number terms, this scenario has an increase in informality from the 787 928 households in 2011 to a peak of 1.34 million in 2017 and then decreasing to a total of 1.05 million households living in informal dwellings (some serviced, and many not). This is significantly less (770 000 less) than that observed in Scenario

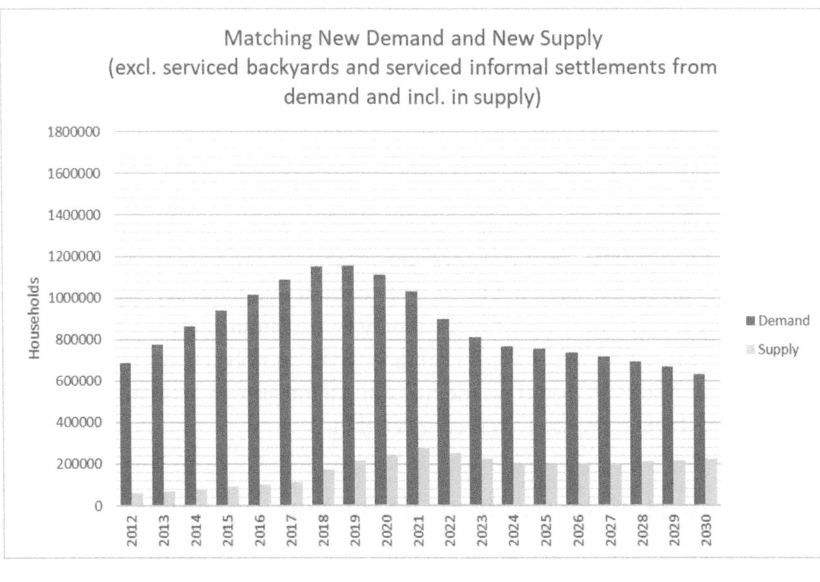

FIGURE 11.7 Scenario 2 showing private sector starting to close the gap between new demand and new supply

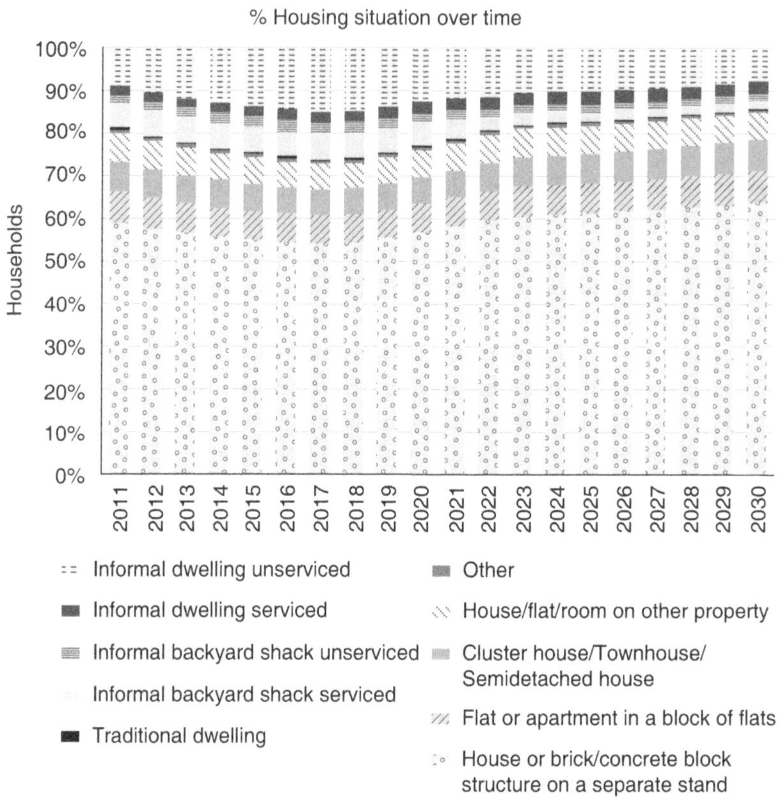

FIGURE 11.8 Scenario 2 showing the Gauteng housing situation over time as private sector supply impacts on growth of informal settlement

1. Figure 11.8 also shows the increasing proportion of households that would have access to permanent housing structures on a separate stand.

By comparing the two scenarios, it is clear that continuing on the pathway of business-as-usual is only going to lead to a rapid increase in the demand for adequate housing, with an increase in informal dwellings up to 26% of the total housing situation by 2030, while the increased private sector scenario can turn this situation around to start reducing informality to a value of 15% by 2030. These preliminary modelling results would suggest that under a medium economic growth forecast the government needs to strongly consider how to bring the private sector on-board and how to leverage their strengths in supporting the housing demand. Bringing this actor on-board will not only be vital ensuring a less informal province but would also aid in the Government meeting its goals and objectives. However, it should be noted that these results are based on the initial medium-growth economic projections for the country, as used when the study was completed in 2016.

4. Possible impact of the pandemic on the housing model

At the time when the model was delivered to the client, in 2016, no one could have foreseen that there would be a pandemic affecting not only South Africa but most countries across the globe. As of 11 April 2021, there have been 417 351 cases of COVID-19 in Gauteng (2.7% of the 15.5 million in Gauteng), with 10 424 confirmed COVID-19 deaths (0.07% of Gauteng's population) and a reported figure of 24 688 excess deaths (0.16% of Gauteng's population) due to natural causes (SAMRC 2021). Hence, unless the pandemic triggers panic, large-scale migration into or out of Gauteng, the pandemic will probably have an insignificant impact on the overall demography in Gauteng.

However, the pandemic is having a significant economic impact. A new scenario was therefore incorporated that takes this economic impact and new forecasts into account: Scenario 3 – COVID-19 impact on the housing sector.

The 'lockdown' effect on unemployment rather than the disease itself has affected the demand for housing as many people have lost their jobs and are unable to afford their current housing situation. There may also have been an impact on housing structure, with many people working from home and children having to move back in with their parents. In-migration trends may also have changed but these and other factors are difficult to predict. The effect on the South African economy is shown in Table 11.2, where the difference between the initial economic growth projections and the current actuals (StatsSA 2020a) and revised growth projections are shown. The revised projections up to 2022 were taken from the SARB MPC forecasts (SARB 2020) up to 2022 and from the Pardee forecasts (Pardee 2020) from 2023 to 2030.

Although the population forecasts remained the same (medium population forecasts), the model shows the significant impact on household income due to the change in the economic forecasts. Figure 11.9(a) shows the number of households per income bracket (where the monthly income displayed represents the upper limit for that income band) for the initial medium-growth scenario, while Figure 11.9(b) shows the households per income bracket for the revised medium-growth scenario.

Figure 11.9(b) actually mimics the overall behaviour in the initial low-growth economic scenario in the model. However, relative to Figure 11.9(a), this second figure indicates that the number of households in the higher income brackets have reduced, with many households not being able to sustain their original high levels of income. There has also been a drop in the R3501-R7501 monthly income band with the income bands below this all increasing

TABLE 11.2 The initial and revised percentage growth figures for GDP used for the model

	2017	*2018*	*2019*	*2020*	*2021*	*2022*	*2023*	*2024*	*2025*	*2026*	*2027*	*2028*	*2029*	*2030*
Initial	3.6	3.6	3.7	3.7	3.7	3.8	3.8	3.8	3.8	3.9	3.9	3.9	3.9	4.0
Revised	1.4	0.8	0.2	-8.2	3.9	2.6	2.0	2.1	2.2	2.2	2.2	2.2	2.2	2.2

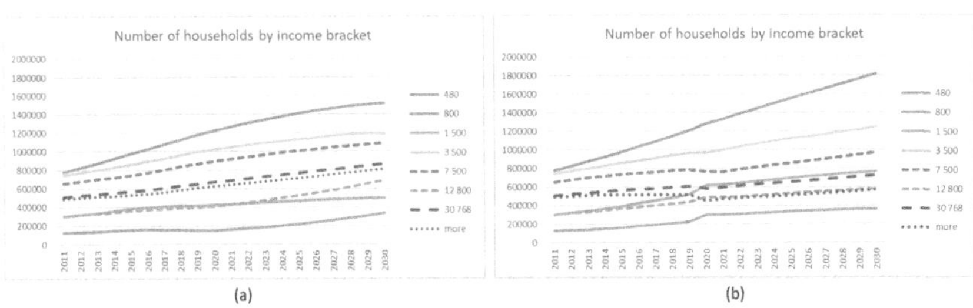

FIGURE 11.9 Number of households by income bracket for (a) initial medium economic-growth forecast and (b) revised medium economic-growth forecast

Another way that government can stimulate production area is through the release of serviced land for housing at greater scale. The ongoing supply of shelter by the informal sector (albeit often inadequate in quality) demonstrates that people can, and will, provide their own houses. What is less feasible without state action is the provision of bulk infrastructure, neighbourhood-level services and amenities, and individual land parcels with legally recognised tenure.

National government and Gauteng as a province are moving towards greater emphasis on the release of serviced land in what is referred to as a rapid land release approach (Makhura 2020; Sisulu 2020). After serviced land is allocated to new owners, this would potentially also open up more space for the private sector (e.g. materials suppliers) to support local and owner-builders in incremental building processes. Support to self-building activities, for example through the people's housing process programme (Napier 2003), can improve the quality of locally built housing, at least for low-rise forms of shelter.

This set of shifts in approach would be more likely to produce a greater volume of adequate stock and that should at least hold out some hope that supply would start to address the new demand arising in the coming years. The value of the modelling exercise comparing the state-dominated business-as-usual approach with the enhanced private sector role scenario is to demonstrate that housing supply would likely improve.

Because over the last two decades, the supply-led approach by the state has set up many path dependencies, the shift towards a greater private sector role will only be realised with bold and concerted action on the part of the state. This does require political will and building mutual trust between the actors.

Using the kind of model presented here, it is possible to start to cost an altered role for the state and the individual levers that it can use to bring about change. A different approach would change the nature of the residential real estate market in the Gauteng City Region over the following few decades. With a wider array of actors and supply channels, greater volumes of housing stock production and a more functional housing market, it is possible that some of the currently unmet areas of demand for adequate housing may then start to be resolved.

6. Conclusions

This chapter has described how the Gauteng Housing Demand Model that emulates housing-choice behaviour based on past trends built on a number of key assumptions, can usefully forecast the possible outcomes of different state housing strategies. After outlining the model methodology, we drew on the data, forecasts and scenarios generated by the model to discuss the alternative futures just outlined.

For government that needs to continuously re-allocate resources and alter strategies and plans to improve outcomes, the practical usefulness and value of this kind of model are substantial.

In this case, the model has been applied to a provincial government in an area of the country that is socially dynamic and economically vibrant. The model accommodated different economic and demographic forecasts. It allowed testing of different mixes of programmatic interventions, and then it effectively helped test different scenarios. In this case, we compared just two possible futures to try to bring into focus the possible outcomes of a greater role for the private sector. When the initial exercise was done in 2016, the economic outlook was more favourable. When re-running the model for post-COVID forecasts (and negative economic growth) the model effectively moved households between the different income bands and proved robust in altering the outlooks in ways that were intuitively convincing. The model also showed how sensitive adequate housing supply is to economic conditions.

To date, the devotion of the South African housing programme to fully subsidised housing products may have caused some to discount the importance of taking measures to improve the ability of households to afford housing.

Being able to model alternative futures ahead of policy change or budget allocation represents a tremendous cost-saving to government. With accurate and up to date data inputs it is possible, to some extent, to avoid having to try out different approaches in practice until the ideal mix of interventions is achieved. Using the modelled scenarios as a basis, it can also open up wider policy debate between parties by showing the likely outcomes of different courses of action.

For future work, there are many possibilities. Particularly important would be to update this and other models with data from the 2021 Census in South Africa. For one thing, this would allow us to look back and assess the accuracy of our previous 2030 predictions when the model was originally applied, and with new data would reveal a more accurate picture for the future. Another area for development would be to extend and enhance the model by undertaking research to provide empirical evidence for the costs and efficacy of some of the recommended private sector demand-side incentives.

Especially with the added challenges arising in society and the economy over the last year, it is very clear from the modelled scenarios that the size of demand for liveable housing will continue to outstrip supply in this part of South Africa unless new, well-thought-out approaches are developed and applied. It has been suggested that this shift will require many different public and private sector actors to focus and collaborate on how best to produce adequate and affordable housing at scale that addresses housing needs across the whole residential market.

Author's note

An earlier version of this research was presented at the African Centre for Cities (ACC) International Urban Conference (Napier 2018).

References

Abramsson M (2012) *Housing Careers*. International Encyclopedia of Housing and Home. Elsevier, pp. 385–389. DOI: 10.1016/B978-0-08-047163-1.00635-4.

Bajari P, Chan P, Krueger D & Miller D (2013) A dynamic model of housing demand: Estimation and policy implications. *International Economic Review*, 52(2), pp. 409–442. www.jstor.org/stable/24517173.

Ballard R & Rubin M (2017) A 'Marshall plan' for human settlements: How megaprojects became South Africa's housing policy. *Transformation: Critical Perspectives on Southern Africa*, 95(1), pp. 1–31.

Boumeester HJFM (2011) *Traditional Housing Demand Research. The Measurement and Analysis of Housing Preference and Choice*. Springer, pp. 27–55. DOI: 10.1007/978-90-481-8894-9_2.

CAHF (2020) *Gauteng Key Indicators 2008 to 2019*. Unpublished report, Centre for Affordable Housing Finance in Africa (CAHF), Johannesburg. http://housingfinanceafrica.org/.

Collier P (July 2014) *Private Sector Crucial to Resolve Housing Challenges*. Private Sector and Development, no. 19. Proparco, Paris, pp. 1–5.

DHS (2004) *"Breaking New Ground" A Comprehensive Plan for the Development of Sustainable Human Settlements*. South African Department of Human Settlements, Pretoria. http://www.dhs.gov.za/sites/default/files/documents/26082014_BNG2004.pdf.

DHS (September 2009) *Housing Project Process Guide*. Department of Human Settlements, Pretoria. www.dhs.gov.za/sites/default/files/documents/publications/housing_project_process_guide.pdf.

DHS (2019) *Department of Human Settlements Annual Report 2018/19*. Department of Human Settlements, Pretoria. www.gov.za/sites/default/files/gcis_document/201911/human-settlements-1819.pdf.

Ferguson B (July 2014) *Housing Public-Private Partnerships in Latin America: Lessons from Mexico and Elsewhere.* Private Sector and Development, no 19. Proparco, Paris, pp. 21–24.

FFC (2012) *Building an Inclusionary Housing Market: Shifting the Paradigm for Housing Delivery in South Africa.* Financial and Fiscal Commission (FFC), Midrand. http://citeseerx.ist.psu.edu/viewdoc/download?doi=10.1.1.361.4659&rep=rep1&type=pdf.

GDHS (12 September 2014) *Gauteng Department of Human Settlements Annual Report 2013/14.* Gauteng Provincial Government, Johannesburg.

GDHS (2015) *Gauteng Department of Human Settlements Strategic Plan 2014/15–2018/19.* Gauteng Provincial Government, Johannesburg. www.gauteng.gov.za/government/departments/human-settlements/Documents/GDHS%20Revised%20APP%202014%20-%202015.pdf.

Hoek-Smit MC & Cirolia L (2019) *Opening-Up the Lower-Middle Income Housing Market in South Africa: The Role of Demand-Side Subsidies.* World Bank Group, Washington, DC.

IHS Markit (2012) *Proprietary City-Wide Demographic and Employment Projections, Company Profile.* IHS Markit. www.ihsmarkit.co.za.

Landman K & Napier M (2010) Waiting for a house or building your own? Reconsidering state provision, aided and unaided self-help in South Africa. *Habitat International*, 34(3):299–305.

le Roux A, Makhanya S, Arnold K, Van Tonder L, Wools L & Mans G (2019) *Green Book – Projecting the Future Growth of South African Settlements: Developing a Concept Model for Town Growth Forecasting in South Africa.* Technical report. CSIR, Pretoria. https://s3-eu-west-1.amazonaws.com/csir-greenbook/resources/WS3_SettlementGrowth_Report_2019.pdf.

Mabin A (July 2013) *The Map of Gauteng: Evolution of a City-Region in Concept and Plan.* Occasional Paper 5. Gauteng City-Region Observatory (GCRO), Johannesburg. ISBN 978-0-620-56079-5.

Makhura D (25 February 2020) *State of the Province Address by Premier David Makhura: Building the Gauteng City Region of our Dreams.* www.gov.za/speeches/state-province-address-premier-david-makhura-25-feb-2020-0000.

Mans GG *et al* (2014) *Background Research Paper for the Integrated Urban Development Framework, Module 1: Demographic Change.* Unpublished research report. CSIR. Pretoria.

Mathibela N (29 January 2019) *FLISP: Answering Your Questions.* Centre for Affordable Housing Finance in Africa (CAHF), Johannesburg. http://housingfinanceafrica.org/documents/flisp-answering-your-questions/.

Mnisi N & Karam A (2020) The impact of the integrated residential development programme on surrounding property values: Case study of Fleurhof, Johannesburg. *Acta Structilia*, 27(1), 2 9–58. www.scielo.org.za/scielo.php?script=sci_arttext&pid=S2415-04872020000100002.

Namponya A & Nkhonjera M (1 October 2020a) *Johannesburg Housing Market Report – 2020.* Centre for Affordable Housing Finance in Africa (CAHF), Johannesburg. http://housingfinanceafrica.org/documents/johannesburg-housing-market-report-2020/.

Namponya A & Nkhonjera M (28 October 2020b) *Ekurhuleni Housing Market Report – 2020.* Centre for Affordable Housing Finance in Africa (CAHF), Johannesburg. http://housingfinanceafrica.org/documents/ekurhuleni-housing-market-report-2020/.

Namponya A & Nkhonjera M (28 October 2020c) *Tshwane Housing Market Report – 2020.* Centre for Affordable Housing Finance in Africa (CAHF), Johannesburg. http://housingfinanceafrica.org/documents/tshwane-housing-market-report-2020/.

Napier M (2003) Supporting the people's housing process. In Khan F & Thring P (eds.), *Housing Policy and Practice in Post-Apartheid South Africa.* Heinemann Publishers, Johannesburg, pp. 321–362.

Napier M (2009) *Making Urban Land Markets Work Better in South African Cities and Towns: Arguing the Basis for Access by the Poor.* Urban Land Markets, Springer, New York, pp. 71–97.

Napier M (1 February 2018) *Housing Demand and Supply in Gauteng through to 2030.* African Centre for Cities (ACC) International Urban Conference, Cape Town, SA.

Napier M, Berrisford S, Kihato CW, McGaffin R & Royston L (2013) *Trading Places: Accessing Land in African Cities.* African Minds, Cape Town.

Napier M & Gavera L (2011) *From Beneficiaries to Citizens: A Discussion Paper.* Urban Land Mark. http://citeseerx.ist.psu.edu/viewdoc/download?doi=10.1.1.361.4061&rep=rep1&type=pdf.

Napier M, Holloway JP, le Roux A & Graham N (2016) *Gauteng Housing Demand and Backlog Findings Report: Phases 3–5, Understanding Housing Demand and Backlog in Gauteng*. Contract report for the Gauteng Department of Human Settlements. CSIR, Pretoria, 73p.

Pain N & Westaway P (1997) Modelling structural change in the UK housing market: A comparison of alternative house price models. *Economic Modelling*, 14(4):587–610.

Pardee (2020) *Economy Forecast for South Africa – GDP*. Pardee Center for International Futures at the University of Denver. www.ifs.du.edu.

Paules CI, Marston HD & Fauci AS (25 February 2020) Coronavirus infections – more than just the common cold. *JAMA*, 323(8):707–708.

Peberdy S, Harrison P & Dinath Y (2017) *Uneven Spaces: Core and Periphery in the Gauteng City-Region*. Research Report 6. Gauteng City-Region Observatory (GCRO), Johannesburg, SA. http://146.141.12.21/handle/10539/23592.

Quantec (2020) *Quantec EasyData Service*. Quantec. www.quantec.co.za/easydata/.

Rust K (2006) *Analysis of South Africa's Housing Sector Performance*. FinMark Trust, 44p. http://citeseerx.ist.psu.edu/viewdoc/download?doi=10.1.1.463.4115&rep=rep1&type=pdf.

SAMRC (2021) *Report on Weekly Deaths in South Africa*. South African Medical Research Council (SAMRC), Bellville. www.samrc.ac.za/reports/report-weekly-deaths-south-africa.

SARB (2020) *QPM forecast Summary Table September 2020 MPC Press Report*. Monetary Policy Committee (MPC), South African Reserve Bank, Pretoria.

Sarda P (July 2014) *A New Generation of Real-Estate Developers Focusing on Low-Income Housing*. Private Sector and Development, no 19, pp. 5–8. Proparco, Paris.

Sisulu L (15 September 2020) *National Minister, 702 Interview*. www.702.co.za/articles/396199/new-strategy-introduced-to-deal-with-housing-backlog.

South Africa (15 November 1994) *White Paper on Reconstruction and Development*. WPJ/1 994, Notice No 1954 of 1994. Parliament of the Republic of South Africa.

South Africa (1997) *Housing Act*. www.gov.za/documents/housing-act.

South Africa (2012) *Constitution of the Republic of South Africa, 1996, as Amended Through to the 17th Amendment*. www.gov.za/documents/constitution-republic-south-africa-1996.

StatsSA (2019) *General Household Survey (GHS) 2018*. Statistical Release P0318. Statistics South Africa, Pretoria. https://www.statssa.gov.za/publications/P0318/P03182018.pdf

StatsSA (2020a) *Gross Domestic Product (GDP), 2nd Quarter 2020*. Statistical Release P0441. Statistics South Africa, Pretoria. http://www.statssa.gov.za/publications/P0441/P04412ndQuarter2020.pdf.

StatsSA (2020b) *Mid-Year Population Estimates, 2020*. Statistical Release P0302. Statistics South Africa, Pretoria. http://www.statssa.gov.za/publications/P0302/P03022020.pdf.

StatsSA (2020c) *Gross Domestic Product, Fourth Quarter 2020: Supporting Spreadsheets*. Statistical Release P0441. Statistics South Africa, Pretoria. http://www.statssa.gov.za/publications/P0441/P04413rdQuarter2021.pdf.

StatsSA (2021) *Stats in Brief, 2021*. Statistics South Africa, Pretoria. http://www.statssa.gov.za/publications/StatsInBrief/StatsInBrief2021.pdf.

Suárez JL, Linage Juan de L & Becerra M (2015) *New Housing Demand Forecast Model*. IESE Research Papers D/1133, IESE Business School, Universidad de Navarra, Barcelona.

Tissington K (2010) A review of housing policy and development in South Africa since 1994. Paper prepared for the Studies in Poverty and Inequality Institute (SPII). http://spii.org.za/wp-content/uploads/2013/12/review-of-the-right-to-housing.pdf.

Van Belle S, Affun-Adegbulu C, Soors W, Srinivas PN, Hegel G, Van Damme W, Saluja D, Abejirinde I, Wouters E, Masquillier C & Tabana H (2020) COVID-19 and informal settlements: An urgent call to rethink urban governance. *International Journal for Equity in Health*, 19(1):1–2. https://equityhealthj.biomedcentral.com/articles/10.1186/s12939-020-01198-0.

Van Niekerk W, le Roux A & Pieterse A (2019) CSIR launches novel online climate risk profiling and adaptation tool: The green book. *South African Journal of Science*, 115(5/6):3. DOI: 10.17159/sajs.2019/6238.

Wang L & Zhang X (2017) System dynamic model for simulating demand-supply interaction of railway express freight system. *Mathematical Problems in Engineering*, 2017, Article ID 8481708.

Zabel JE (2004) The demand for housing services. *Journal of Housing Economics*, 13(1):16–35.

12

APPRAISAL OF THE NIGERIAN REAL ESTATE MARKET

Constraints and opportunities

Austin C. Otegbulu and Victor A. Akujuru

Introduction

The real property market is simply the arrangement by which buyers and sellers of various types of land use – virgin land, agricultural estate offices, industrial buildings, shops, retail store – are brought together to determine a price at which a particular property can be exchanged. Real property market is an abstract term aggregating all transactions in real property throughout the country. In this respect, it is possible to distinguish sub-markets for different types of property; for example prime shop properties for different geographical locations, office blocks, residential homes, etc. The market has no formal organization or central agency or institution like the stock exchange (with the exception of property companies listed in the stock exchange). The property market is also imperfect (Balchin & Kieve 1985; Ling & Acher 2010). The preceding explanation shows how real estate is distinct from other markets. According to Ling and Acher (2010), real estate is the single largest component of wealth in our society. In Nigeria, it is the most attractive investment though highly constrained by poor mortgage financing. Because of its magnitude, it plays a key role in shaping the economic condition of individuals, families and friends. It can substantially influence a family's ability to finance its education, health care and other important needs. Hence, changes in the value of real estate can dramatically affect the wealth of business and the capacity to grow. Similarly, real estate resources can greatly affect a community's ability to attract and support profitable business activities as well as provide secured, convenient and affordable living environment for the citizens. The adequacy of public infrastructure, including roads, bridges, dams, airports, schools and parks affects the quality of life and the performance of the property market. Many writers are of the view that real estate represents about 50% of the world's wealth. The real estate market is influenced by various macro and micro factors. The ability to identify indicators that could be used to measure changes in real estate market will be of great use to investors and other stakeholders in making informed investment decisions with regard to real estate.

The property market is influenced by social, economic and physical/environmental forces which ultimately drive demand and supply in a market area. A good understanding of the

DOI: 10.1201/9780429279256-12

market dynamics operating in a particular market is critical to real estate investment decision. Real estate market is important for each country not only because it provides building structures and infrastructure necessary for life and work but because it has a strong impact on the development of the whole nation due to its multi-economic influence (Car 2009). Real estate market analysis provides guidance for decision making in real estate investment and the foundation of most property professional opinion. The property professional is mostly employed for his understanding of the property market condition. The professional is conversant with the demand and supply position at each point in time and hence is in a position to make appropriate judgement that will guide informed investment decisions based on market information. There is no doubt that the Nigeria property market is highly under-reported, and this has created a lot of misconception and ignorance at the level of market activities. As a matter of fact, the flow of international investment activity has not been without direction. The direction of flows without doubt is set by investor's judgment on the existence of quality markets. In the globalized investment world, market competitiveness is defined solely by investor perception, relative endowment with infrastructure and related urban services and a favorable institutional environment. It is for this reason that markets are analyzed in the investor/participants context along with the opportunities and associated operational structures that enable and support particular forms of investment (Osuntogun 2005; Mcreal et al. 2002; cited in Dugeri 2011). The property market is a major sector and good attraction for direct foreign investment. It is also a major contributor to economic growth and GDP.

With huge housing deficit and high demand for real estate space and assets in different sub-markets, great potential and opportunities exist in the Nigeria property market. This potential and opportunities are however constrained by poor infrastructure and urban services, political instability and ever fluctuating policies, irregular economic and property cycle, high lending and fluctuating exchange rate among others.

With rising rents, huge housing deficit, poor institutions and over regulation, the Nigerian property market is a matrix of opportunities and unique challenges which only well-informed and optimistic investors would be able to successfully convert into viable long-term benefits.

The aim of this study is to evaluate the challenges and opportunities that exist in the Nigeria property market.

This will be achieved with the following objectives:

- To review the current market situation.
- To determine existing opportunities in the Nigerian property market.
- To examine the challenges of infrastructure to property market development.
- To determine the effect of fluctuating monetary exchange rate on market trend.
- To evaluate other institutional related challenges as constraints to the property market.

Real estate is one of the best money making sectors in the world, but investors rely on detailed market information and analysis before investing their money. This information is lacking in most African and third world countries, hence the need for this report.

Review of the Nigerian property market, and opportunities

The Nigerian economic cycles have been very irregular moving in tandem with fluctuating oil prices. The economic and property cycles are like Siamese twins. They cannot be separated. There is therefore the need for real estate market analysts to understand the dynamics of market cycles and how these cycles affect real estate demand and supply and value estimate. Real estate markets operate through the dynamic interaction of supply and demand which can be perceived as cycles of activity and real estate value is likely to fluctuate during different phases of a cycle, so analysts must understand and address these cycles (Fanning 2005).

This is because a developer who understands the nature of the property cycles will be in a better position to plan a project ensuring as much as possible that it is completed and released into the market at the most appropriate time. This will help in minimizing risk and maximizing investment returns. A lot of developers in Nigeria don't make use of professionals in their real estate investment decision and hence rely on their personal experience and intuition to make decisions involving huge capital outlay. They consult professional appraisers only when they are compelled by their financiers to do so. In this case they are only interested in getting the loan irrespective of whether the project will be viable or not. It is important that developers engage a property market research consultant or assemble an in-house professional research team. According to Reed and Sims (2014), a smart developer will always carry out his own research to establish the impacts on their business and adjust their development strategy accordingly. One common problem in the study of property cycle is to assume that property fluctuation dynamics is uniform in a city. Different parts of a city may experience different patterns in their cycle. In Lagos and other Nigerian cities, property values in some areas might be rising while declining in other areas. During economic recession property values in upper end locations like Ikoyi and Victoria Island (both commercial and residential) were experiencing voids and declining rental while the middle and lower ends were more stabilized probably due to a filtering down process. This was observed during the recessions of 2004, 2016 and 2020 in Nigeria.

A study carried out by Chukwu et al (2016) in two neighbourhoods of Enugu South East of Nigeria indicated that rental values of properties in the new Haven area were increasing while those of Achara Layout were decreasing due to differentials in the quality of infrastructure in both areas. The same scenario can be witnessed in Akure South west of Nigeria (See Iroham et al. 2013).

Nigeria is a victim of mono-economy depending mainly on revenue from crude oil exportation. It has experienced periods of boom and downturn, thus influencing the economic cycle. After a boom period marked by high oil prices and double digit-growth, Nigerian real estate sector was significantly impacted by the country's economic downturn with growth declining in 2015 and 2016 and had remained subdued in 2017, 2018 and 2019.

This has slowed down or stalled some high-end office and residential projects, making developers increasingly turn their attention to smaller and more affordable projects. This is noticeable at Eko Atlantic where many high rise commercial and residential projects have been stalled by poor demand and low absorption rate; also in Ikoyi and Victoria Island where the pace of construction has drastically gone down without a known completion date. These include: Westhold Mall, Victory Mall, Royal Garden Mall, Falomo Plaza etc. The Nigeria's real estate trajectory has risen and fallen with the international oil market influencing the economic cycles. According to local real estate consultancy firm Estate Intel (citing data from National Bureau of Statistics (NBS)), real estate growth rate moved from 15.1% in the first quarter of 2013, peaking at 24.14% in the second quarter of 2013 with full year real GDP growth hitting 20.49% in 2013. As oil prices began to slide in mid-2014, the nominal term rate of 22.4% in the first quarter of 2014 dropped to 10% in the second quarter of 2014 with full year growth hitting 12.5% in 2014. Nominal growth fell to 9.18% in the third quarter of 2015 and 8.27% in the last quarter of 2015 with full year growth standing at 9.52% in 2015 before dropping sharply to 0.16% in the first quarter of 2016 as the price of Brent crude oil sunk to less than $30 per barrel. According to real estate consultancy MCORE, the real estate service sector contracted by 8.38% in nominal terms in the first quarter of 2018 which is 18.94% less than the growth rate in the first quarter of 2017. The sector in real terms contracted by 9.4% in the first quarter of 2018, against 7.67% in the fourth quarter of 2017. In real terms, the sector contracted by 9.4% in the first quarter of 2018, a 3.48% decline from the first quarter of 2017. The sector contracted 5.63% to real GDP growth in the first quarter of 2018, against 6.34% in the first quarter of 2017 and 7.03% in the fourth quarter of 2017. Northcourt Real Estate, a real estate firm based in Lagos, in its 2018 real estate market outlook, reported that the 2012–2014 boom period had brought residential real estate to perhaps national level with the 2016 recession offering necessary market corrections with price, yields and other performance indicators corrected in 2017. Residential vacancy rate averaged 11% in 2017, down from 15.57% in the first half of 2017 and 32.87% at the end of 2016. The vacancy rate in Abuja averaged 7% in 2017 and 25.57% at the end of 2016. The Northcourt report also indicated that developers are resuming work on developments that stalled during the recession, while residential supply in the Government Reserved Area (GRA) Ikeja began to decline partly because some buildings were converted into office spaces due to increased demand in that sector. Quest for joint venture developments was increased in Lagos and other major cities of the country to enhance property value and overcome the problem of development finance. In the Abuja property market, the rate of property development outpaced infrastructure provision resulting in many districts having developments without urban services.

The office property market is yet to recover from the after effect of the post 2015 economic recession. It has experienced more voids than the residential sector. In Lagos, the green property market is emerging with four major developments completed. These are Nestoil building (9,904sqm), Heritage place (15,600sqm), the wings (27,000sqm) and Alliance Place (7,000sqm). The first three are LEED green certified while Alliance Place is a World Bank certified green building. The rental value ranks between $500 and $750 per square metre. Green buildings are expected to have higher rent cost and lower operational cost which should result in low service charge. This is not fully the case in the Lagos green building market. The service charges for green certified buildings are far higher than that of the non-certified green buildings like the Landmark place, NIPOST/Energy building, and Civic Centre. This could be due to high void rate in the buildings forcing few tenants to share the burden of common services. According to Otegbulu (2017), the sustained collapse in the oil prices, falling household income, dollar shortage and weakening naira has strangled the demand for both commercial and residential real estate in Lagos and other parts of the country. Despite the dwindling economy, investors are not pulling out their investment because they have faith in the long-term prospect of the Lagos real estate market. According to a 2017 property market by Uson Eleh & Co (a Lagos based estate surveying firm), emerging markets funds particularly from South Africa, familiar with African markets, are deploying funds to acquire interest in large prime development projects which are expected to yield high returns in a market that is still very much under supplied or unsaturated. According to the same report, real estate in 2016 saw low transaction levels from the investment side as a result of declining economic condition. The waiting period for both space and assets remained on the increase. Feelers from stake holders in the real estate market indicated a 40% void and rent fall in prime office space and 20%–30% in prime residential properties. This is a result of poor economic conditions in the country. The government announced that the country has recovered from the economic recession, but its effect is yet to be felt in all sectors of the economy. Management surveyors in the past have devised means of keeping their tenants from moving out of their prime properties. Under this situation, good and promising tenants with good prospect of bouncing back to good business are encouraged to stay by giving them attractive incentives. Their rent (Head rent) which might be $400 per square metre may be retained, but they are allowed a rent free period of three months annually for a period of between two and four years when the economy is expected to bounce back. This provides a better option because if the tenants move out, the property may remain void for many years. Other incentives may include fit out allowance (rent free for a period allowed for office partitioning and furnishing) and service charge discount until a new tenant moves in. In the lower end of the property market, properties are experiencing stability or slight increase in rent particularly for incoming tenants.

Importation and production has declined due to downward slide of the economy and an increase in the exchange rate of the dollar and decline in bank lending and import financing. The direct effect of this is decline in demand for warehouses and related properties with subsequent increase in void level and fall in rental value. This view is also reinforced by the 2017 property market report of Ubosi Eleh & Co and the views of major players in the market.

Back to the office market, the introduction of new spaces into the market is growing. MCORE forecast that about 64,000sqm of office space will be in the Lagos market before 2020. These include: Alliance Place, Madina Towers with 8,300sqm, Kingsway Towers with 13,317sqm and Cornerstone Towers with 12,000sqm of gross lettable area (GLA). Most of these are already in the market as the developments were all completed. For properties whose rents are dollarized, landlords are inclined to reduce rent and also accept rent in local currency to remain in business. There is no doubt that the real estate market of any country or city is linked to both micro- and macro-economy variables. The most critical symptoms of an economic recession are negative economic growth, decreasing production and growing unemployment. The implication of this is decline in the economy including falling prices of homes and real estate in general. This supports the views of Geogievski (2015) that in financial crisis, the real estate market is the most sensitive. It is the first to respond when there is expansion or decline in the economy. According to the 2019 real estate market report by Ubosi Eleh & Co, despite being in recession, Nigeria's real estate sector showed signs of exiting contraction mode in the third quarter of 2018. The growth performance reported for the quarter was linked to Nigeria's economic performance and readjustment in the country's property market, government and private sector fund injection and the concluded general elections of 2019 (according to the figures posted by the National Bureau of Statistics). At the end of the 9th month of 2018, the real estate sector posted a growth of -2.68% in contrast to the -3.88% in QI of the same year. It is important to note that in the past few years the supply and demand for co-working office space have increased. Its main attraction is that it provides enormous cost saving by doing away with initial set-up costs, hidden charges in conventional leases. Savings of up to 70% of monthly cost can be achieved. In addition, facilities such as break-out areas, meeting rooms and video conferencing systems are also available to businesses without additional costs to the investments resulting in the substantial savings and reduced upfront costs. It is also more flexible than the traditional office space (Ubosi Eleh & Co 2019); this is an innovation in the Nigeria office property market. There is also significant growth in the retail space market as more malls are being developed in different parts of the country. Most of these malls, if not all, have Shoprite as their main anchor tenant.

Lagos remains the most attractive city in terms of real estate investment. In the height of this attraction new projects are still ongoing though at a slower pace. The largest being the Eko Atlantic City, a public and private partnership initiative between Lagos state government and a private company, South Energy Nigeria Ltd (a subsidiary of Chagoury group of companies).

The city is to cover 10million square metres of land adjacent to Victoria Island and provide homes for 250,000 people. Nigeria's large population and growing economy provides enormous opportunities for investors both local and foreign. The market is growing and housing deficit is 17–20 million.

Nigeria's property market, particularly that of Lagos, has been improving in terms of transparency and ease of doing business, according to Jones Lang Lassales (JLL) Global Estate Transparency Index (2018). In 2018, Nigeria ranked 67th among 100 markets, up 16 notches of deviation from 83rd in 2016. The federal government of Nigeria set up a committee to improve ease of doing business in Nigeria, and this appears to be yielding results.

Opportunities in the real estate market

A lot of opportunities exist in the Nigeria property market. It however requires creativity and prudent planning to tap into it. The Nigerian economy is about the biggest in Africa, and this makes it attractive to investors in spite of its existing challenges. This will be enhanced if there is a more vibrant and borrower friendly mortgage system with single digit interest rates. This is in addition to removing unnecessary and choking bottlenecks in land registration system. A lot of opportunities exist in the following areas of the property market.

Apartment hotels (also known as short lets)

There is a high demand for this type of property particularly for interstate and international travelers, and those who need a place to operate for a short period. It is furnished like a normal home and can be shared by families and friends. It is also ideal for short-term lodging and office use. The demand for such is at its peak during festive periods.

Co-work office space

The demand for co-working has been on the increase in recent times, even though it is constrained by the COVID-19 pandemic. A lot of business people have discovered that they have no need for a permanent office space but a place they can go and work or hold meetings as the need arises. It is trending, and the demand for it was on the increase before the pandemic.

Joint venture

The problem of a poor mortgage system in Nigeria has created a good opportunity for joint ventures and partnership as land owners can partner with developers to develop their land under terms and conditions to be agreed upon by both parties.

Opportunities for institutional investors

Real estate provides a wide range of opportunities for institutional investors like pension fund, insurance companies etc. It is also easier for institutional investors to create alternative investment vehicles for real estate or go public and use an equity fund for real estate development. Currently, UAC property development company UPDC Plc is the only active public quoted real estate company in the Nigerian Stock Exchange. Almost all the development projects embarked on by UPDC are successful. A lot of opportunities exist for institutional real estate developers in the Nigerian capital market, including the opportunity to finance real estate development. There is high demand for mass residential development, and they are in the best position to fill the gap.

High demand for residential real estate and shops

There is an undersupply or deficit in the housing sector. For more than two decades, the gap has been speculated at 17 million units but a realistic estimate should be about 30 million or more. The undersupply is more on the middle and lower end of the residential property market. In most Nigerian cities demand for shopping space is high and commands very high rental value. In Lagos Island, a shop of 4 square metres could command a rent as high as N300,000–N400,000 (about $1,000) per annum. There is a scarcity of good shopping space.

Challenges of Nigeria property market

The Nigeria property market is a matrix of prospects and challenges. The challenges are daunting and include infrastructure, finance, institutional and regulatory framework, exchange rate, high cost of construction, political instability, insecurity etc.

Infrastructure

Infrastructure is a major constraint in the Nigerian property market as real estate is diminished in utility without adequate and sustainable infrastructure. Infrastructure serves as the spinal cord that links people, social institutions, socio economic activities and natural environment into a coherent urban relationship. It is a key component for providing an enabling environment for sustainable growth. It is equally essential for safeguarding health, protecting the environment and promoting the efficient operation of human settlements (Otegbulu, 2014). In spite of the importance of infrastructure in human settlements, most developers have to provide roads, drainage and electricity to access their site in addition to providing estate roads, drainage and other related infrastructure.

This in no small measure escalates cost of construction and constrains housing affordability.

Financing

In civilized and advanced countries individuals and institutional developers rely on mortgage to finance their real estate projects, but in Nigeria the reverse is the case. Apart from a scarcity of mortgage funds the lending rate is quite high and discouraging. It could be as high as

25%. Very few have access to mortgage financing. Only about 5% of the 137 million housing units in Nigeria are financed with mortgages; the mortgage debt to GDP ratio is about 1% while mortgage accounts for less than 1% of commercial banks total assets (specialist mortgage banks account for 5% share of the market CBN 2019). Loans for home construction are more common though still insignificant at 1.7% of total loan disbursement (IMF 2019).

Title registration

In his book *The Mystery of Capital*, Hernando De Soto (2000) argued that capitalism triumphs in the west and fails everywhere else because of the challenges of registration of land title in the developing countries. He demonstrated in practice that titling hitherto untitled assets is an extremely effective way to promote economic development. He indicated that in some countries it took several administrative steps (207 steps) and several years (six years) in 52 government offices and for title registration, and 728 steps to obtain building permit.

This is cumbersome and could be discouraging to both local and foreign investors. In Nigeria, property registration is expensive, while land titles are prone to political interference because of the vast powers of the state governors under the land use act. The governor can revoke title for political reason under the guise of over-riding public interest. Nigeria is notoriously one of the worst in the global context when it comes to registering property according to the World Bank ease of Doing Business 2019 report. Nigeria ranks 184th out of 190 countries (World Bank 2019). In Lagos, the registration can last an average of 105 days and an average of12 procedures (steps) and costs about 11.1% of the value of the property. With regard to construction permits, Nigeria still lags behind and ranks 149th out of 190 countries according to the same report (World Bank 2019). Ghana requires six procedures (steps), 47 days and 6.1% of the property value. The cost of property registration in Nigeria is about 60% higher than sub-Saharan Africa and almost three times the QECD average. Without reliable title documentation it will be impossible to access a mortgage loan either to build or buy a property. The issue of land titling is a major constraint in the Nigerian property market.

Unstable foreign exchange

Nigerian currency is very unstable and oscillates like a pendulum. It is very weak against the US and other currencies like the dollar or pound sterling etc. In consequence this leads to high cost of construction particularly in buildings with very high foreign content. The exchange was 148 naira to one US dollar in 2010 and 360 naira to one US dollar in 2019. It affects the real value of real estate and distorts real estate investment forecasting. For this reason, rent for some upper end properties is charged in US dollars to hedge the rents against the fluctuating value of the naira, but some landlords were forced to reduce their dollar rent due to economic challenges in the country.

Property market data

There is paucity of property market data, and most transactions are shrouded in secrecy. There is need for more research and standardization of market operation. Robust market data should be developed and reviewed periodically. Most times available data are obsolete and misleading thus constituting disincentives to investors and financiers. The lack of data also affects the advisory services of real estate investment advisers and valuers in the country.

Terrorist activity and insurgency

Nigeria is not free from terrorist activists and insurgency, including kidnapping, and this in no small measure affects the property market in the concerned states. Terrorist activities, insurgency and Muslim–Christian clashes have drastically affected the property market in Northern Nigeria. The hot spots include Maiduguri, the capital of Borno state. Many residents of state capitals Yobe and Adamawa have fled to other parts of Northern Nigeria including Abuja and parts of southern Nigeria particularly Lagos for safety and continuation of their economic life. The direct consequence of this is increasing voids and vacancy rate in the hot spot areas and increasing demand in the safe areas. According to stakeholders the vacancy/void rate in the affected areas is about 30%–40%.

Impact of COVID-19 pandemic

The impact of the COVID-19 pandemic on property demand is more on commercial properties as it made many organizations force about 70% to 90% of their staff to work from home. This is likely to continue in the near future. Towards the end of 2020, most organizations gave notice of intention to reduce their current office space by the end of the year. This will increase the incidence of void in office buildings. E-commerce is also on the increase and will decrease demand for shopping space.

Conclusion

A review of the Nigerian property market indicates that it is a matrix of challenges and opportunities. The property cycle operates in sync with the economic cycle and moves in a harmonic manner. The gestation of the property cycle is distorted by scarcity of property space and assets which leads to faster recovery in spite of lingering economic stagnation and unemployment. The challenges include poor infrastructure, poor mortgage finance opportunities, high interest rate, high cost of construction, cumbersome process and high cost of land title registration, terrorism and insecurity in parts of northern Nigeria, poor market information, unstable exchange rate and low level of direct foreign investment in real estate due to poor market environment and cloudy market information among others. There is a need to ease and standardize the process of title registration and development permits, encourage local production of building materials, improve the ease of doing business and liberation of mortgage lending institutions and process to enhance market activities. There is no doubt that Nigeria's property market is shrouded in a sea of opportunities.

References

Balchin, N. P. and Kieve, J. L. (1985) *Urban Land Economics*. Macmillan Publishers Ltd, London.

Car, L. M. (2009) Selection of factors influencing the residential property prices in Slovakia. *Journal of Current Issues*, vol. 17, no. 3.

Central Bank of Nigeria (CBN) (2019) GDP growth in 2019. https://af.renters.com/article/investingNews/illAFRCNIR210F- OZABS.

Chukwu, A. C., Aniagolu, C. O. and Obodo, C. M. (2016) Trends in rental value of residential properties in Enugu, Nigeria; A comparative study between New Haven and Achara layout. *Journal of Multidisciplinary Engineering Science and Technology,* vol. 3, no. 2, pp. 4037–4047.

De Soto, H. (2000) *The Mystery of Capital.* Basic Books, New York, NY.

Doing Business (2019) World Bank. www.worldbank.org/content/dam/doingbusiness/media/Annual-Reports/English/ DB2019 – report-webversion.pdf.

Dugeri, T. M. (2011) An evaluation of the maturity of the Nigerian property market. Unpublished PhD thesis submitted to the Department of Estate management, University of Lagos, Lagos.

Fanning, S. F. (2005) *Market Analysis for Real Estate.* Appraisal Institute, Chicago.

Geogievski, B. (2015) The impact of Greece financial crisis on Balkan real estate maket. *Journal of Developing Country Studies*, vol. 5, no. 2.

IMF (2019) *International Monetary Fund Annual* Report. IMF, Washington, DC.

Iroham, C. O., Olawunmi, A. O., Simon, R. F. and Akerele, B. A. (2013) Assessing the trend in rental value of property along Oyemukun road Akure Nigeria. *Journal of Research in the Buett Environment (Covenant University Otta, Nigeria)*, vol. 1, no. 1.

Jones Langs Lassale (JLL) (2018) Global real estate transparency index 2018. https://www.jll.co.za

Ling, D. C. and Archer, W. R. (2010) *Real Estate Principle.* McGraw-Hill Publishers, New York.

Nigeria (2018) Housing finance Africa. https://housingfinanceafrica,org/app/uploads/Nigeria-2018-finalpdf.

Otegbulu, A. C. (2014) Urban infrastructure condition and neighbourhood sustainability: A contingent valuation approach. *Ethiopian Journal of Environmental Studies and Management*, vol. 7, no. 2, pp. 160–170.

Otegbulu, A. C. (2017) An evaluation of Lagos property market. AFRES Conference Johannesburg South Africa.

Property in Nigeria (Nigeria property centre). www.nigeriapropertycentre.com/properties?Limitstart towers.

Reed, R. and Sims, S. (2015) *Property Development.* Routledge, London.

Ubosi Eleh & Co (2019) The Nigeria real estate report 2019. https://www.ubosieleh.com https://drive.google.com/drive/1/0/folders/ifi

World Bank (2019) *Ease of Doing Business Report.* Washington, DC: The World Bank.

13

CHALLENGES OF RAISING REAL ESTATE FINANCE IN NIGERIA

Akinola. E. Olawore and Austin C. Otegbulu

Introduction

Real estate development is a multifaceted business encompassing various activities ranging from construction of new buildings, renovation and re-lease of existing buildings to the purchase of land and the sale of improved parcel to others. Real estate is contained in everything we buy, sell or use. It is a component of cars, watches, television, clothing, furniture etc. Real estate is a necessary and useful component of all activities carried out by human beings – either individually or in organizations. The ability of a property to provide services needed by individuals and organizations is termed productivity and because real estate is productive, people often need to buy and sell it (Smith & Corgel, 1987). By implication, real estate has utility and value and hence can be traded in the markets and held as an investment. Finance is very critical to real estate investment. Real estate is capital intensive requiring a huge sum of money to be expended to actualize its creation or acquisition. Real estate finance is described as the totality of funds required for the execution of all the operations or activities required for the development of real estate and its acquisition.

The best real estate development and investment project ideas will amount to nothing more than wishful thinking unless there is capital in place to finance or fund the project. It is also important to note that finding sources of capital for funding both land acquisition and development can be daunting as sources may shift in sync with dynamics of economic conditions and regulations. The real estate financing system is under-developed in Nigerian economy and this is compounded by the absence of relevant long-term capital in the financial market. The performance of any real estate finance system will always depend on the volume and nature of funds within the economy and the proportion of it that can be spread, mobilized or even dedicated for real estate. When considering real estate finance, it should be noted that investments in real estate is ever competing with other classes of investments like stocks and shares, gold, artwork, government bonds, treasury bills etc. for the investible funds within a given economy. For this reason, investors continue to migrate from the stock market to the real estate market, hence there is a greater need for sound financial analysis of income producing properties than ever before Berges (2004).

DOI: 10.1201/9780429279256-13

In Nigeria, most developers rely on either equity or debt (mortgage) finance or both. However, the mortgage industry in Nigeria is currently untapped and relatively unstructured, and underfunded. These in effect have a negative impact on real estate development and the property market. Financing the purchase of real estate usually involves borrowing on a long- or short-term basis. Considering the huge amount involved in real estate development, the prices paid for real estate, financing costs are usually significant in amount and weigh heavily on the decision to buy property as investors must understand how these costs are computed and how various provisions in loan agreement affect financing costs and mortgage payment (Brueggeman & Fisher, 2011). Most borrowers are unmindful of the cost implication of the mortgage loan and in some situations abandon the project when they find out the loan cannot be repaid from the real estate income. However, most developers build to sell (dealer developers) to avoid accumulation of interest payment and profit loss.

The Problem

The problem of real estate development within the Nigerian economy is daunting and this is complicated by lack of funding. In the past decade or more, Nigerian housing deficit is said to be in the region of 17 million. One can safely assume that the figure on account of rapid urbanization and slowed paced of property development activity if extrapolated must have reached 25 million or more. The mortgage industry in Nigeria is currently underfunded and relatively unstructured. According to a World Bank Report cited by Usman (2014) about 80% of the population live in informal housing structure of varying degree of permanence and with no ownership right, and the country requires about 375 million USD to bridge the housing deficit. About 70% of this deficit is concentrated in Lagos, Port-Harcourt and Kano. The problem of informal settlement and difficulty in obtaining title registration to land constrain the ability of citizens, corporate and individual, to access mortgage finance, thus creating billions of dead capital according to De Soto (2000). The problem of finance is not restricted to the housing sector but affects the other sectors of real estate like offices, retail shops, hotels etc. The problem of unreliable market data/information and opaqueness in property market activities dissuades foreign investors coupled with the unstable exchange rate of the naira against the US dollar. The instability in exchange rate is also a major constraint to reliable real estate investment income forecasting. This challenge is further compounded by bubble and bursts in the market triggered by market dynamics (property and economic cycles). The Nigerian real estate market has a lot of potential but is currently only contributing about 4% to the country's GDP when it has the potential to contribute in double digits to the country's GDP if properly structured and funded. Another major problem of real estate finance is the high cost of construction and high interest rate. According to Ogedengbe and Adesope (2017) the problem of real estate financing centres on high interest and inflationary rates in the Nigerian economy.

Literature

Real estate in whichever form, be it residential, commercial, industrial, agricultural, etc., plays a critical role in the life of citizens and the economic development of a nation and contributes significantly to the GDP of individual countries. It is estimated that real estate constitutes about 70% of Nigeria's wealth. In spite of this, real estate financing in Nigeria is fraught with many challenges. Financing of real estate is both from the formal and informal sectors. Currently mortgage financing is underdeveloped hence the real estate sector is not mortgage driven but informally driven. Direct Foreign Investments (DFI) is on the low side while development financing from capital market sources is still emerging. In the residential sector, scarcity is more on the lower end property market compared to the upper end. One can easily find a lot of vacant upper end properties including commercial properties. In developed economies, mortgage finance plays a pivotal role in the development of property markets. The spill-over effect of strong property markets in the form of job creation, particularly in the construction sector, improvement in living conditions, potential development of long-term finance market for infrastructure development, freeing up of government resources to meet other socio-economic needs and ultimately in improved economic growth cannot be underestimated (Rama et al. 2013).

According to Ezimuo et al. (2014), the well-established and tested methods of real estate financing are: equity capital, loan capital, mortgage funding, debenture and contractor financing. In addition to the preceding, funding real estate through capital market sources is becoming popular in developing economies and is considered a good panacea to the problem of high interest rates. There is actually not much difference between loan capital and mortgage finance as methods of real estate finance. For the avoidance of doubt, a mortgage is simply a pledge of a property to secure a loan. It is not a promise to pay anything. It is the promissory note, the actual promise to pay, which accompanies or in some cases, becomes a part of the mortgage instrument, which is the proof of debt (Smith (2013). It is therefore doubtful if a housing loan not tied to the property or where a property is not pledged as security against the loan could be described as a mortgage. In most cases, the loan is secured on the employee's salary.

Real Estate Financing Sources

As earlier mentioned in this chapter, real estate development is capital intensive and may not be possible to be wholly fund from personal savings. Due to this, developers all over the world, including Nigeria, resort to different sources of finance to fund real estate. In Nigeria, the common sources of real estate funding are indicated next:

Commercial Banks

Commercial banks in Nigeria grant loans for the purpose of financing real estate. Commercial bank loans are usually short-term mortgage, bridging finance or in very rare cases long-term tenured. Commercial banks in Nigeria rarely give long-term loans for real estate development or acquisition. Some big commercial banks have a real estate loan or mortgage department or unit. In some situations, they act as guarantors for foreign loans. Since the commercial banks are more inclined to short-term loans in the nature of their business structure and regulatory requirements, most real estate developers are more dealer developers than investment developers. This implies that they build to sell off rather than retain as an investment for future income generation. If the developers build as an investment to pay back the debt from rental income, it is doubtful if the rental income can sustain periodic capital and interest repayment as the interest rate on mortgage and other loans has remained as high as 25% per annum for over a decade.

Mortgage Banks

In Nigeria, the mortgage banking system is structured and hinged on three structures: the Federal Mortgage Bank, National Housing Fund and primary mortgage institutions. In 1956, the Nigerian Building Society (NBS) was established to provide mortgage loans. During the same year, the colonial government introduced the African Staff Housing Fund, which was established to assist civil servants to own their own houses (Otegbulu, 2000). The NBS was a joint venture of the Commonwealth Development Corporation and the federal and Eastern Government of Nigeria, but with the promulgation of the Indigenization Act 1973, the federal government acquired 100% ownership of the NBS and in effect renamed it as Federal Mortgage Bank of Nigeria (FMBN). The Mortgage Institutions Act of 1989 accorded the Federal Mortgage Bank the apex mortgage bank status as a regulator of mortgage institutions. The FMBN was again restructured in 1994 following the promulgation of the FMBN Act of 1993 and saddled with the role of managing the National Housing Fund and unbundled of its core mortgage operations. The Federal Mortgage Finance Limited (FMFL) was subsequently established in 1993 to carry out the retail aspect of mortgage lending hitherto performed by the FMBN. The FMFL is expected to provide long-term credit facilities to primary mortgage institutions. The Federal Mortgage Bank is expected to coordinate mortgage lending in Nigeria using resources from deposits mobilized and equity contribution by the federal government and Central Bank of Nigeria (CBN) (Elisha, 2013). Other primary mortgage institutions were licensed to mobilize capital savings from the public and grant housing loans to individuals.

The National Housing Fund (NHF) program created in 1989 as part of the federal government of Nigeria's strategy to address mobilizing the problem of large pool of loanable funds at noncommercial rates was accorded legal backing in 1992; it raised a lot of hope as many believed it will ease the problem of housing finance for workers. It was established pursuant to the National Housing Fund Act (1992) with the aim of mobilizing funds through compulsory monthly contributions by public service workers for the provision of affordable housing. The fund is expected to guarantee a more consistent source of funding through membership contribution. At the turn of 2005, the total loan amount disbursed was in the sum of N10 billion with just above N7 billion loaned to real estate developers. The impact of FMBN was largely insignificant in the early years of its existence (Rama et al. 2013). From 2002 to 2012 the FMBN approved loans worth N184.6 billion for the provision of 57,206 housing units. Out of this amount, only the sum of N94.8 billion was disbursed. The situation has not changed much in 2020 as the housing deficit has remained on the increase. According to Rama et al. (2013), the bank also participated in secondary mortgage and capital mortgage operations. This is evidenced in the issuance of the first tranche of its N100 billion FMBN Mortgage-Backed Securities (MBS) bonds in 2007 to enable civil servants to buy houses during the federal government monetization policy.

In spite of all these efforts, the mortgage system is underdeveloped and funds less than 5% of real estate developments in Nigeria. The federal government in furtherance of its objective in sustaining the availability of long-term funds through the capital market has caused the establishment of the Nigerian Mortgage Refinance Company (NMRC) which as made mortgage bond issuances. As at 2021, the NMRC has N29 billion from the capital market, and has attracted $120 million from multilateral International Development Association (IDA) loans (NMRC, 2021).

Equity Finance Sources

Equity funds are mainly generated by the individual or organization for the purpose of funding real estate developments. It could come from salaries, business projects, savings, contributions etc. Amongst these, the chief source of equity funding is savings. This is for both individuals and corporate organizations. According to Soyendo (2015), equity sources could be private or public. Private equity comes from individual or corporate savings; that is, retained earnings, asset stripping, cash or revenue reserve of companies over a period of time and accumulated savings of individuals, friends and thrift. Public equity is also another source and is derived from invitation extended to the public to subscribe equities or ownership in a real estate company. In Nigeria, there are very few public quoted real estate companies. Only one of them is active in the capital market, and that is UACN Property Plc. This is an area highly unexploited for real estate development considering the high lending rate in commercial banks in addition to the absence of long-term loans. In the Nigerian property market, more than 70% of development comes from equity funding at times from questionable sources also.

Forward Sale/Off Plan

An extension of equity funding is forward sale or off plan sales of a property. It is the sale of a property to a given buyer or buyers on a particular date with a promise to deliver at a future date, and the agreement so entered binds the developer to deliver on the same date. It is becoming popular among developers in Nigeria as a result of the absence of a well-developed mortgage system and high interest rate. In the light of the global financial crisis tightening its grip on diverse sectors of the economy, many private developers are turning their backs on commercial banks for real estate development loan. In addition, the scissors effect of commercial loan constrains the viability of loan funded property development. Banks are also reluctant to fund real estate development due to a glut in the property market. Bank loan to real estate has been on the decline since 2016. As at first quarter of 2019, a total sum of N593.3 billion was allocated to the sector which is 4% lower than N615.3 billion allocated in the first quarter of 2015. Data or information from the National Bureau of Statistics (NBS) shows that percentage of non-performing loans rose by N33.31billion increasing from N102.74 billion at the end of 2018 to N136.05 billion in 2019.

According to *The Guardian* newspaper of 6th July 2020 (Page 27) most real estate developers complain of the rigorous process of obtaining commercial loans for real estate development and high lending rate. Most developers resort to off plan or forward sale and partnership funding with interested parties as a more viable option compared with a bank loan. The development companies hire marketers who they train in marketing their proposed project though off plan sales. The investors are presented with 3D drawings and pictures of the proposed developments and are encouraged to pay in instalments, which makes acquisition more convenient.

Some development companies float an investment company under their group with the sole aim of attracting funds from the public with the promise to pay them 35%50% of interest of their investment on the maturity date which is usually 12 months from the date of investment. The money so invested is reinvested in a property development and other viable businesses. The interest payable looks attractive enough that most investors may decide to ignore inherent risks associated with high investment return.

Real Estate Investment Trusts (REITs)

A real estate investment trust is a company that owns and, in most cases, operates income-producing real estate. REITs own many types of commercial real estate, ranging from office and apartment buildings to warehouses, hospitals, shopping centres, hotels and commercial buildings. They are corporations or trusts that use the pool capital of many investors to acquire and manage income property and/or mortgage loans. They are traded in the Nigeria Stock Exchange (NSE) like stocks; hence, you can buy and sell REITs through your stock broker as with other type of shares.

There are three major real estate investment trusts in Nigeria: UPDC Real Estate Investment Trust, Sky Shelter Fund and Union Homes Real Estate Investment Trust. UPDC is the largest of the REITS by asset holding. As at 28th July 2018, it has an asset value of N33.2 billion followed by Union Homes at N9 billion.

The UPDC REIT holds prime real estate investments such as Victoria Mall, Plaza 1 & 2 and Abebe courts. Others include UAC Complex Abuja and a warehouse in Aba, Abia State of Nigeria. It is listed on the floor of the Nigerian Stock Exchange. Union Homes Real Estate Investment Trust also has some key investments, namely: Victoria Court Ikoyi, Olives apartment Jabi Abuja, and Locke Apartments Lekki, Lagos. It is worthy to note that the impact of REITs in the Nigeria property market leaves much to be desired. According to Victor Gbonegun (9th March 2020) reporting in *The Guardian* newspaper, the impact of real estate investment trusts as an alternative home finance naturally designed to reduce the 17 million housing deficit in the country has left much to be desired over time, with little or no significant impact. According to the same writer, the return on investment in Nigeria is below that of other investors in other economies. The return on REITs in Nigeria is at 7% compared with 15% in South Africa and 9% in Kenya. The gap is glaring. There is therefore the need to improve on the REITs market performance and management to make it attractive to investors. In addition, poor familiarity of the Nigerian Investment class of REITs as an asset class has also constrained its growth. There is also the need for the media and Nigerian Institution of Estate Surveyors and Valuers (NIESV) to promote investment in REITs as an asset class and investment option.

Constraints to Real Estate Funding in Nigeria

Real estate financing in Nigeria is fraught with many challenges which need a well-articulated policy mix of programmes to address. Notable among the factors militating against sustainable real estate financing include the following:

Poor Mortgage Structure

The mortgage market in Nigeria is poorly structured, and sub-optimally funded. This limits the quantum of mortgage loan to be advanced to borrowers for real estate development and investment. The capacity of primary mortgage institutions to fund big projects is highly limited for the simple reason that these institutions are also under capitalized.

Low Level of Awareness Among Citizens of Mortgage Facilities

The use of mortgage as a means of real estate financing is not very popular amongst Nigerians. The few who are aware are scared to approach relevant financial institutions due to delay in concluding mortgage loan transactions, and short tenure of most mortgage loan advances, in addition to high lending rates.

High Interest Rates Charged by Domestic Financial Institutions

The interest rate charged by commercial banks in Nigeria for real estate development is scary. In most cases, it ranges between 18% and 25% compared with 3%–5% in developed countries (See Rama et al. 2013). For this reason, most Nigerians prefer to rely on personal savings and traditional sources like contributions from associations. The high lending rates suffocate developments and constrain their viability.

Low Awareness of Capital Market Opportunities

There is generally a low level of awareness of the opportunities in the capital market for real estate development. These include equity shares, REITs, unitization securitization etc. These are major sources of real estate finance and refinancing in the developed countries. There is need to reposition and popularize the capital market for real estate financing.

Difficulty in Title Registration

In its 2012 doing business report, the World Bank ranked Nigeria as 180th in a pool of 183 economies on the ease of registering property. The ease of doing business in land registration is discussed with regard to procedure, time and cost to transfer a property and the quality of the administration system. According to the World Bank release, Doing Business 2020, transferring property in Nigeria requires on average 12 procedures and transaction costs to be incurred more than 15% of the property value. According to the said report, Nigeria improved considerably as it moved by 38 spots in the global rankings from 169th in 2017 to 131st 2020. This notwithstanding it is still difficult to register property in Nigeria as only 3% of land in Nigeria is registered, creating an abundance of dead capital on land. You cannot approach any bank for mortgage without a valid title on your property. Title to property is a major constraint to mortgage lending and has encouraged the promotion of informal driven property market.

Conclusion

Finance is a major issue in all aspects of real estate development/property market: residential, industrial, commercial etc. According to a recent World Bank report, about 80% of Nigerians live in informal housing structures with varying degree of permanence with no ownership right and in effect, constrained from accessing a formal source of finance for real estate development. The problem of high interest rate remains a nightmare. There is an urgent need to restructure and reposition the real estate finance infrastructure to create a more vibrant and formal property market. The capital market needs to be more vibrant and reorganized to play a major role in real estate finance, followed with wide publicity to create awareness. There is also a need to improve the process of land registration, reduce mortgage lending rates and improve on the capacity of mortgage and commercial banks to provide loan advance for real estate development. Real estate is very central to the wealth of any nation and the government in association with the private sector need to review the existing real estate financing structure and infrastructure to enhance real estate development in the country.

References

Berges, S. (2004) *Complete guide to real estate finance for investment properties.* John Wiley & Sons Inc.

Brueggeman, W.B., and Fisher, J.D. (2011) *Real estate finance and investments.* McGraw Hill.

De Soto, H. (2000) *The mystery of capital.* Basic Books.

Elisha, J.D. (2013) Mortgage financing in Nigeria. Central Bank of Nigeria Occasional paper no 50. CBN, Abuja, Nigeria.

Ezimuo, P.N., Onyejiaka, C.J., and Emoh, F.I. (2014) Sources of real estate finance and their impact on property development in Nigeria: A case study of mortgage institutions in Lagos metropolis. *British Journal of Environmental Research*, vol. 2, no. 2, pp. 35–58, June.

Gbonegun, V. (2020) Why real estate investment trust penetration remains low in Nigeria. *The Guardian*, March 9.

NMRC (2021) Success stories. www.nmrc.gov,ng on April 16.

Ogendegbe, P.S., and Adesope, A.A. (2017) Problems of financing real estate in Nigeria. *Journal of Human Ecology*, vol. 14, no. 6.

Otegbulu, A.C. (2000) Housing for all in the 21st century: Developing private sector initiative. Urban Behavior and Property No 9, Vol. 1. Department of Geography University of Lagos.

Rama, U., Yakubu, J., Bewaji, P., Adigun, M.A., and Elisha, J.A. (2013) Mortgage financing in Nigeria Central Bank of Nigeria (CBN) occasional Paper No 50. CBN Abuja.

Smith, I.O. (2013) *Practical approach to land of real property in Nigeria.* Ecowatch Publication.

Smith, H.C., and Corgel, J.B. (1987). *Real estate perspectives.* Richard D. Irwin

Soyendo, V. (2015) Real estate finance and taxation. Dissertation Submitted to College of Engineering, Design Art and Technology for the Award of BSc Land Economics, Makere University.

Usman, H.M. (2014) *Nigerian real estate development financing opportunities Nigeria real estate industry overview.* Aso Savings. https://pafcoffice.com.

14

PERFORMANCE OF SECTOR-SPECIFIC AND DIVERSIFIED REITS IN SOUTH AFRICA

Omokolade Akinsomi

1. Introduction

Real estate investment trusts (REITs) are highly sought after assets by investors globally and in Africa, and this is because the underlying assets of REITs are income-producing properties, which are generally stable, and REITs distribute substantial amount of their profits to shareholders as dividends. On the African continent, the South African listed REITs market is the most active and capitalised with a current market capitalisation of R193 billion (US$19.35 billion) as at December 2019.

South African REITs are regulated by the Johannesburg Stock exchange (JSE) and are expected to comply with a number of criteria in respect to listing requirements as stipulated by the JSE. These include REITs must own a minimum of 300 million Rand (US$46.8 million)[1] of property, 75% of revenue must be generated from property related incomes as well as indirect property investments, debt level cannot exceed 60% of gross asset value and would need to pay a minimum of 75% of taxable earnings to shareholders each year.

Table 14.1 shows the rank of South Africa in the FTSE EPRA NAREIT global indices as at August 2020. The index series represents a trend in REITs globally. South Africa remains the only REITs market in Africa included in the index. Out of 37 REITs markets globally listed on the index, South Africa ranks number 21 with a weight of 0.30% contribution to the index and a market capitalisation of US$4.76 billion.

South African Real Estate Investment Trusts (SA REITs) is considered as a leading listed property investment vehicle in Africa as well as globally. Figure 14.1 shows that South African REITs grew in the number of funds from 17 REITs in April 2013 to 27 REITs as at September 2020. In terms of market capitalisation, SA REITs grew from US$18.7 billion in April 2013 to a record high of US$30 billion in September 2018; the current market capitalisation of SA REITs as at Septembers 2020 is US$8.4 billion and has been adversely affected by the covid-19 pandemic. SA REITs have witnessed a 230% reduction in market capitalisation from US$19.35 billion in December 2019 to US$8.4 billion as at September 2020. With 27 REITs listed on the Johannesburg stock exchange and market capitalisation of US8.4 billion, SA REITs are ranked at number 1 in Africa and number 21 globally based on EPRA FTSE NAREIT indices.

DOI: 10.1201/9780429279256-14

TABLE 14.1 FTSE EPRA NAREIT Global Indices

Rank	*Country*	*Market Capitalisation (USDM)*	*No. of Constituents*	*Weight* (%)
1	USA	752,405	122	47.81
2	Japan	164,136	48	10.43
3	China	115,895	58	7.36
4	Germany	87,831	11	5.58
5	Hong Kong	85,238	12	5.42
6	United Kingdom	69,039	39	4.39
7	Singapore	47,463	17	3.02
8	Australia	47,351	12	3.01
9	Canada	39,384	21	2.50
10	Sweden	27,630	14	1.76
21	South Africa	4,761	9	0.30

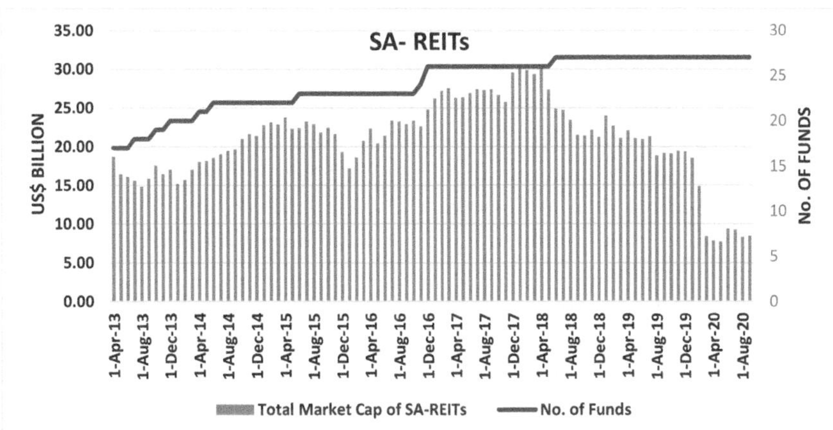

FIGURE 14.1 Growth in Market Capitalisation for SA REITs: April 2013 to September 2020

Source: Authors' compilation from IRESS Research Domain

As seen in Figure 14.2, diversified REITs appear to be the more dominant REITs structure in South Africa. The market capitalisation of diversified REITs in April 2013 was US$13.04 billion compared to the market capitalisation of specialised REITs at US$5.62 billion. Diversified REITs grew in market capitalisation to a high of US$20.27 billion in September 2018 and are currently valued at US$5.41 billion in comparison to the market capitalisation of sector-specific REITs at US$3.01 billion. The South African REITs industry in recent years and months has been affected by the South African economy[2] as well as the covid-19 pandemic. This observation in South Africa differs for instance in countries such as Australia where sector-specific REITs are more dominant than diversified REITs. Lin et al. (2019) document that sector-specific REITs represented 90% of the total size of the Australian REITs as at 2018 over the past 18 years.

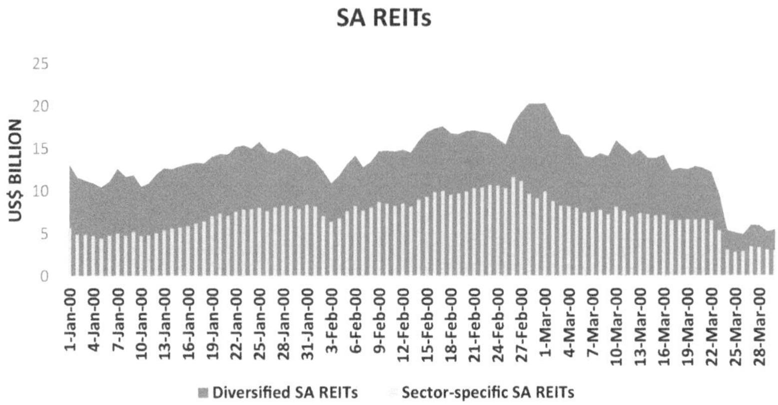

FIGURE 14.2 Growth in Market Capitalisation of SA REITs: April 2013 to September 2020

Source: Authors' compilation from IRESS Research Domain

Past studies have examined what type of REITs (diversified and sector-specific REITs) perform better. For instance, while authors such as Lin et al. (2019) find that sector-specific REITs outperform diversified REITs in Australia, other authors such as Benefield et al. (2009) examine US REITs and find that diversified REITs perform better than sector-specific REITs. Authors such as Gyourko and Nelling (2006) reveal that sector-specific REITs are more risky than diversified REITs and that diversified REITs do not show any evidence of outperformance.

From a South African perspective, while authors such as Ntuli and Akinsomi (2017) have examined the initial performance of SA REITs and Akinsomi et al. (2017) explore the role of gold market speculation in herding of SA REITs, there is a lack of work done to examine the performance of diversified REITs and sector-specific REITs in South Africa. This chapter examines the performance of the diversified and sector-specific REITs in South Africa from April 2013 to September 2020. The chapter examines risk-return performance employed risk-adjusted measures, Sharpe ratio and the capital asset pricing model to examine the performance of diversified and sector-specific REITs in South Africa. Since we find that diversified REITs are the most dominant in South Africa, we ask a research question (RQ):

RQ: Do diversified SA REITs offer a superior risk-adjusted return compared with sector-specific SA REITs?

2. Understanding South African REITs

As seen in table 14.2, diversified REITs dominate the SA REITs landscape with 50% of REITs domiciled in South Africa considered diversified according to the IRESS categorisation. Diversified REITs rank as number 1 in South Africa valued at US$5.42 billion, followed by retail REITs in number 2 valued at US$1.87 billion. Industrial and office ranks at number

TABLE 14.2 Profile of Sector-specific SA REITs. September 2020

*Sectors**	*No. of Funds*	*Market Cap (US$ Billion)*	*% of SA REITs*
Diversified	14	5.42	50%
Industrial and Office	2	0.64	7%
Residential	2	0.11	7%
Retail	8	1.87	29%
Specialty	2	0.39	7%
Total	28	8.43	100%

Source: Authors* compilation from IRESS Research Domain (2020)

*Categorised by IRESS Research Domain

TABLE 14.3 Profile of Leading SA REITs. September 2020

Rank	*SA REITs*	*Listed Date*	*Sector*	*No. of Properties*	*Market Cap (US$ Billion)*
1	Growthpoint	November 1987	Diversified	568	2.21
2	Fortress	October 2009	Diversified	284	0.93
3	Resilient	December 2002	Retail	38	0.88
4	Redefine	October 2013	Diversified	133	0.83
5	Equities	June 2014	Industrial/Office	62	0.63
6	Investec	April 2011	Diversified	98	0.44
7	Stor-age	November 2015	Specialty	79	0.31
8	Vukile	June 2004	Retail	76	0.28
9	Liberty	November 2018	Diversified	16	0.27
10	Hyprop	1988	Retail	20	0.25

Source: Authors* compilation from IRESS Research Domain (2020) *Categorised by IRESS Research Domain

3 with a market capitalisation of US$0.64 billion whilst specialty and residential rank at number 4 and 5 respectively with a market capitalisation of US$0.39 billion and US$0.11 billion respectively.

Table 14.3 shows the leading REITs in South Africa as at September 2020: 50% of the leading REITs in South Africa are diversified REITs, 30% of the leading REITs in South Africa are retail REITs, while 10% are industrial and office REITs and specialty REITs. Growthpoint REIT is the leading South African REIT valued at US$2.21 billion with 568 properties, followed by Fortress REIT valued at US$0.93 billion, Resilient – a retail REIT – valued at US$0.88 billion, Redefine Properties ranks at number 4 with a market capitalisation of US$0.83 billion. Equities REITs is categorised under industrial and office valued at US$0.63 billion, followed by Investec (US$0.44 billion). Stor-age, a specialty REIT, ranks at number 7 valued at US$0.31 billion, followed by Vukile at US40.28 billion, Liberty 2 Degrees valued at US$0.27 billion and at rank 10 is Hyprop, a retail REITs with 20 properties under management with a market capitalisation valued at US$0.25 billion.

This chapter contributes to the real estate literature in a number of ways. First, this study is the first to examine if a value-add exists in diversified REITs in South Africa since this structure of REITs is the most dominant. Second, diversified REITs are significant in the South Africa REITs landscape, however there is a lack of understanding if this domination leads to superior returns. This study therefore seeks to investigate this. The finding of this study is significant for investors in order to make informed decisions on potential REITs investments; the study also seeks to assist managers of REITs in making informed decisions in terms of portfolio allocation.

Section 3 reviews existing literature on the performance of REITs, section 4 describes the data and the methodology employed, section 5 analyses the results and section 6 concludes.

3. Literature Review

There are a number of literature on the sector performance of REITs; however, there is no consensus on which performs better (sector-specific or diversified REITs) and this remains a debated issue with results showing superior performance for both types of REITs.

Ro and Ziobrowski (2011) examine if specialised REITs outperform diversified REITs from 1997 to 2006; the authors employ a capital asset pricing model (CAPM) and Fama-French 3 factor model. The authors do not find any evidence that REITs that specialise in specific properties outperform their diversified counterparts. Their results show that sector-specific REITs are riskier than diversified REITs and that diversified REITs do outperform sector-specific REITs although their results were not statistically significant. A study investigating Australian REITs by Lin et al. (2019) finds that sector-specific REITs from January 2000 to August 2018 have better risk-adjusted returns, increased portfolio returns and enhanced portfolio diversification compared to diversified REITs in Australia.

Benefield et al. (2009) examine whether property-type diversified REITs outperform property-type sector-specific REITs. Their findings suggest superior performance of property-type diversified REITs; the authors employ Treynor index, Sharpe ratio, single-factor and 3-factor model to analyse the data from 1995 to 2006. Anderson et al. (2015) further examine the role of property type diversification on REITs. The authors find that diversified REITs perform better than specialised REITs in the areas of higher cash flows, return on assets and performance.

Gyourko and Nelling (1996) in earlier studies on sector vs diversified REITs examine property type holdings and location of investments of REITs. Their findings reveal that retail-centric REITs exhibit more beta risk in comparison to industrial REITs and that diversification across property markets shows no evidence of any meaningful diversification.

Some authors have examined the role of sector and diversified performance from a portfolio construction perspective. Moss et al. (2017) investigate the performance of different portfolios of REITs that specialise in particular properties compared to a diversified free float market capitalisation index. They find that a two-step process, which involves a combination of REITs with a trend following overlay, can offer benefits compared to a passive benchmark index.

Whilst some authors have examined the roles of sub-sector REITs in a mixed asset portfolio, Newell and Wen (2006) investigate the role of non-traditional real estate sectors in REITs portfolios from 1994 to 2005; their findings show that non-traditional REITs such as self-storage REITs were the second best performing in a mixed asset portfolio, with the best performing being a more tradition REITs in industrial REITs. Cho (2017) explores Japanese REITs from 2010 to 2015 and the diversification benefits of sector-specific REITs and their role in a mixed-asset portfolio. Results in this study show that hotel and industrial REITs provide the best risk-adjusted returns among sub-sector REITs during the period investigated.

From a South African perspective, authors such as Ntuli and Akinsomi (2017) examine the initial performance of REITs from 2013 to 2015 and find that REITs in a mixed asset portfolio were return enhancing. Akinsomi et al. (2017) investigate the role of gold market speculation in explaining REITs returns, and results show that between 2008 and 2011 speculation in the gold markets contributes to herding in the REITs industry in South Africa.

Existing literature on South African REITs on the performance of sector-specific REITs in comparison to diversified REITs is scant and in some cases non-existent. Therefore, this study attempts to provide answers to which REITs investment strategy outperforms specifically from an emerging market perspective such as South Africa, which has received less attention.

4. Data and Methodology

We examine South African REITs from the period April 2013 to September 2020. The study follows the classification of diversified REITs and sector-specific REITs by IRESS Research Domain, a leading financial database in South Africa. The monthly REITs stock prices, 3-month Treasury bill rate, 10-year bond rates and JSE indexed is retrieved from the IRESS research domain.

To analyse the performance of the SA REITs, over the period April 2013 to September 2020, monthly risk-adjusted returns of SA REITs of diversified REITs and sector-specific REITs are investigated including Sharpe ratios as well as the analysis of risks (beta) and outperformance (alpha) using a capital asset pricing model.

Table 14.3 lists the profile of the REITs listed on the Johannesburg stock exchange domiciled in South Africa. There are 36 listed REITs on the Johannesburg Stock Exchange, 28 of these REITs domiciled in South Africa are seen in table 14.4.

Eight of the REITs listed such as Capital & Counties Properties PLC, RDI REIT, Schroeder European Real Estate Investment Trust, Investec Australia Property ltd, Stenprop Limited, BK One Ltd, Capital & Regional PLC, Hammerson PLC are domiciled outside South Africa in countries such as United Kingdom and Australia. As seen in table 14.4, there are 14 active diversified REITs, two active industrial and office REITs, two active residential REITs, eight active retail REITs and two active specialty REITs.

Employing monthly returns of SA REITs, a market capitalisation REITs index of sub-sectors was constructed as seen in figure 14.3 from April 2013 to September 2020 which includes diversified, industrial and office, residential, retail and specialty REITs. All REITs sub-sector begin at 100 in April 2013 except residential REITs which begins June 2015 due to the JSE listing of Indupalce. Generally the REITs indexes performed well between 2015 and 2018, and poor performance was recorded in 2019 and worse off in 2020 as a result of the covid-19 pandemic.

TABLE 14.4 List of Active Domiciled REITs Listed on the JSE (April 2013 to September 2020

Active SA REITs
Diversified REITs
1. Arrowhead Prop LTD A
2. Attacq
3. Dipula Income Fund A
4. Emira Property Fund Ltd
5. Fortress REIT ltd A
6. Growthpoint Properties
7. Investec Property Fund ltd
8. Liberty Two Degrees Ltd
9. Rebosis Property Fund Ltd
10. Redefine Properties Ltd
11. SA Corporate Real Estate Fund
12. Spear REIT ltd
13. Texton Property Fund ltd
14. Tower Property Fund limited
Industrial and Office REITs
15. Delta Property Fund Ltd
16. Equities Property Fund ltd
Residential REITs
17. Indluplace Properties ltd
18. Transcend Residential Prop
Retail REITs
19. Accelerate Property Fund
20. Exemplar Retail ltd
21. Fairvest Property Holdings ltd
22. Hyprop Investments
23. Octodec Investments ltd
24. Resilient REIT ltd
25. Safari Investments ltd
26. Vukile Property ltd
Specialty REITs
27. Hospitality Property Fund ltd
28. Stor-age Property REIT ltd

Source: Authors'* compilation from IRESS Research Domain *Categorised by IRESS Research Domain

To analyse the performance of diversified and specialised REITs in South Africa from the period April 2013 to September 2020, we employ average annual returns, average annual risk and Sharpe ratio. Results are available in table 14.5 and formulas as seen in the following:

$$Average\ Annual\ Returns_{monthly} = Average \frac{P_0 - P_{-1}}{P_{-1}} \qquad \text{(Equation 1)}$$

$$where\ P_0 = Closing\ price\ on\ the\ last\ day\ of\ the\ months,$$
$$P_{-1} = Closing\ price\ on\ last\ day\ of\ previous\ month$$

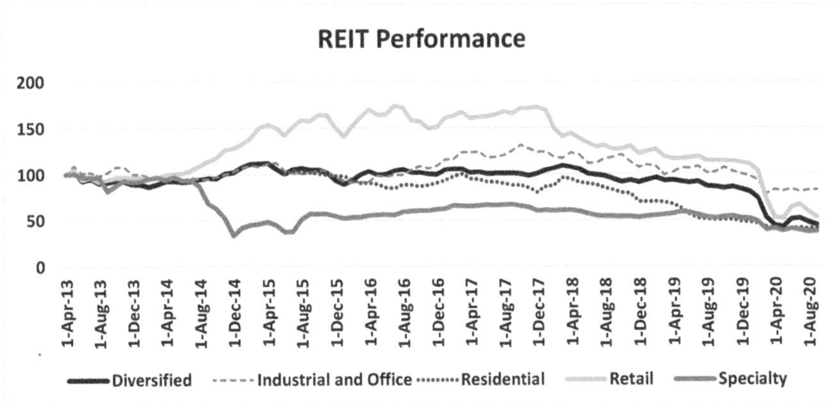

FIGURE 14.3 Sub-Sector SA REITs Market Capitalisation Index: April 2013 to September 2020 (100 = April 2013)

Source: Authors' compilation/analysis

$$Average\ Annual\ Risk_{monthly} = Average\sqrt{\frac{\sum_{i=1}^{n} X_i - \bar{X}}{n}} \qquad \text{(Equation 2)}$$

$$where\ X_i = Closing\ price\ on\ the\ last\ day\ of\ the\ month,$$
$$\bar{X} = average\ annual\ closing\ price\ on\ last\ day\ of\ the\ 12\ months$$

$$Sharpe\ Ratio = \frac{\in\left[R_a - R_{fr}\right]}{\sigma_a} \qquad \text{(Equation 3)}$$

$$where\ R_a = Asset\ Return,\ R_{fr} = Risk - free\ rate,$$
$$\sigma_a = standard\ deviation\ of\ the\ excess\ return\ of\ asset\ a$$

$$Treynor\ Ratio = \frac{\left[R_a - R_{fr}\right]}{\beta} \qquad \text{(Equation 4)}$$

$$where\ R_a = Asset\ Return,\ R_{fr} = Risk - free\ rate,\ \beta - beta\ of\ the\ portfolio$$

To estimate beta and alpha (outperformance) in table 14.6, we use the capital asset model as seen next:

$$\in R_i = \alpha + \beta_i\left(R_m - R_{rf}\right) + e \qquad \text{(Equation 5)}$$

$$where\ R_i = Asset\ Return,\ \alpha = \text{alpha or putrperformance},\ R_m = \text{expected return of the market},$$
$$R_{fr} = Risk - free\ rate,\ e = error\ term$$

5. Analysis and Results

Risk-Adjusted Returns

In other to answer our posited research question-, do diversified SA REITs offer a superior risk-adjusted return compared with sector-specific SA REITs? As seen in table 14.5, we examine the average annual return, average annual risk and Sharpe ratio and compare the estimates across diversified and specialised SA REITs. Stocks had the highest average annual returns in South Africa from April 2013 to September 2020, this is followed by industrial and office REITs (−2.66%), retail REITs (−3.66%), specialty REITs (−5.99%), diversified REITs (−8.78%) and residential REITs (−11.32%). In terms of the performance indicators in the form of average annual returns, diversified REITs ranked number 5 in comparison to the stock market and other specialised REITs. Residential REITs are the only specialised REITs with lower returns than diversified REITs during the period investigated. In terms of risk level, stocks shown the lowest risk level during the period investigated with 2.13%, followed by office and industrial (4.81%), residential (5.08%), diversified (5.79%), specialty (6.29%) and the highest risk level are by retail REITs at 9.82%. Finally Sharpe ratio of each asset class is compared, this allows comparison and an understanding of returns compared to its risk level. Stock ranked number one in South Africa with a Sharpe ratio of −1.93, followed by retail REITs with −4.55, specialty with 5.19, industrial and office with −5.65, diversified in 4th place at −5.65 and residential REITs ranked lowest in terms of performance measured by the Sharpe ratio at −8.53.

Over the period investigated, the results provide strong evidence that specialised REITs outperform diversified REITs from 2013 to 2020. Our results are in line with the findings of Lin et al. (2019) who investigate Australian REITs and find that sector-specific REITs perform better than diversified REITs from the period January 2000 to August 2018.

Diversification Benefits

In table 14.6, the study examines the diversification benefits of diversified, specialised REITs and stocks during the period April 2013 to September 2020. In comparison to stocks, diversified REITs are the most correlated with general stocks at 0.5600, followed by retail REITs correlated with stocks at 0.4786, industrial and office correlated at 0.4678, followed by specialty REITs correlated at 0.2277, whilst residential REITs are the least correlated with stocks at 0.1985. The results of the correlation analysis suggest that specialised REITs have

TABLE 14.5 Sector Specific SA REITs adjusted return analysis: April 2013 to September 2020

Asset Class	*Average Annual Return*	*Average Annual Risk*	*Sharpe Ratio*	*Rank*
Diversified	−8.78%	5.79%	−6.13	5
Industrial and Office	−2.66%	4.81%	−5.65	4
Residential	−11.32%	5.08%	−8.53	6
Retail	−3.66%	9.82%	−4.55	2
Specialty	−5.19%	6.29%	−4.59	3
Stocks	4.85%	2.13%	−1.93	1

Author's compilation/analysis

TABLE 14.6 Sector Specific SA REITs correlations analysis: April 2013 to September 2020

Asset Class	*Stocks*	*Diversified*	*Industrial and Office*	*Residential*	*Retail*	*Specialty*
Stocks	1					
Diversified	0.5600***	1	1	1	1	1
Industrial and Office	0.4678***	0.4370***	0.1927	0.2532**	0.1854*	
Residential	0.1985	0.4071***	0.3987***	0.2474**		
Retail	0.4786***	0.8893***	0.2189**			
Specialty	0.2277**	0.2456**				

*Note:****, ** and * denotes significance at the 1%, 5% and 10% level respectively
Author's compilation/analysis

TABLE 14.7 Risk Adjusted Performance Measures: April 2013 to September 2020

	Diversified	*Industrial and office*	*Residential*	*Retail*	*Specialty*
Panel A: risk	0.7985***	0.5300***	0.1896	0.8480***	0.4896**
Beta					
Panel A: risk					
Adjusted performance	–0.3508	–0.4649	–1.6666	–0.2905	–0.5531
Treynor Ratio	–0.0156***	–0.0138***	–0.0319***	–0.01203*	–0.0209**
Alpha (Outperformance)					

*Note:****, ** and * denotes significance at the 1%, 5% and 10% level respectively
Author's compilation/analysis

higher diversification benefits with general stocks in comparison to diversified REITs as they were less correlated with general stocks. Diversified REITs had the highest correlation coefficients with general stocks during the period investigated. Lin et al. (2019) in line with our results find that sector-specific REITs provide diversifications benefits in comparison to diversified REITs.

Table 14.7 shows the risk adjusted performance measures of diversified and specialised REITS from April 2013 to September 2020. The Treynor ratio and alpha (outperformance) are measures employed for comparative analysis of returns while the beta is an estimate of risk, the estimates are employed for further analysis on the performance of specialised and diversified REITs. The Treynor ratio measures how much risk adjusted return was generated for each unit of risk taken. Retail REITs has the highest Treynor ratio at −0.2905, followed by diversified REITs at −0.3508, industrial and office REITs at −0.4649, specialty REITs at -0.5531, residential REITs has the least Treynor ratio during the period at −1.666 from April 2013 to September 2020. The analysis examine the alpha (outperformance) of each SA REITs in comparison to the market index, retail REITs outperformed other sector REITs in South Africa at −0.01203, followed by industrial and office at -0.0138, diversified REITs at −0.0156, specialty REITs in fourth place at −0.0209 and the least performed REITs during the period measured by alpha (outperformance) is residential REITs. In terms of risk measurement, beta is employed. The least risky SA REITs during the period investigated from April 2013 to September 2020 is residential REITs with a beta of 0.1896. This is followed by specialty REITs with a beta of 0.4896, industrial and office with a beta of 0.5300, diversified REITs have the second highest risk measure with a beta of 0.7985. Retail REITs have the highest beta at 0.8480 during the period investigated in this study.

6. Conclusion

Prior works have examined the performance of specialised and diversified REITs, and results are mixed at best and inconclusive. A number of past literature make an argument that diversified REITs outperform specialised REITs Benefield et al. (2009), Ro and Ziobrowski (2011). Lin et al. (2019) examine Australian REITs and find that specialised REITs outperform diversified REITs. We therefore investigate the performance of specialised and diversified REITs for the first time in Africa by examining South African REITs during the period April 2013 to September 2020.

Our findings from an SA REITs perspective show that specialised REITs have superior average returns, Sharpe ratio and Treynor ratio which are all measures of outperformance in comparison to diversified REITs. Further analysis employed to measure returns is alpha (outperformance) – specialised REITs outperform diversified REITs during the period investigated from April 2013 to September 2020. Our findings are in line with Lin et al. (2019) who examine Australian REITs and find similar results in favour of specialised REITs.

These findings are of particular importance for individuals, investors and portfolio managers in an emerging market such as South Africa, which has received little attention in the performance of specialised and diversified REITs. The findings of this work would assist in the understanding of the performance of REITs as well as portfolio allocation considerations by REITs as well as fund managers in choosing the appropriate REITs within their investment portfolio. Our findings reveal that sector-specific REITs are more desirable in South Africa within the period investigated.

Notes

1 The exchange rate as at 9th November 2020 was US$1 to R15.59c.
2 The GDP of the South African economy fell by 51% in the second quarter of 2020. Prior to this GDP of South Africa has declined for the fourth consecutive time since the second quarter of 2019.

References

Akinsomi, O., Balcilar, M., Demirer, R. and Gupta, R., 2017. The effect of gold market speculation on REIT returns in South Africa: A behavioral perspective. *Journal of Economics and Finance*, 41(4), pp. 774–793.

Anderson, R.I., Benefield, J.D. and Hurst, M.E., 2015. Property-type diversification and REIT performance: An analysis of operating performance and abnormal returns. *Journal of Economics and Finance*, 39(1), pp. 48–74.

Benefield, J.D., Anderson, R.I. and Zumpano, L.V., 2009. Performance differences in property-type diversified versus specialized real estate investment trusts (REITs). *Review of Financial Economics*, 18(2), pp. 70–79.

Cho, H., 2017. The role of Sub-sector REITs in mixed-asset portfolios in Japan. *Journal of Real Estate Literature*, 25(2), pp. 389–408.

Gyourko, J. and Nelling, E., 1996. Systematic risk and diversification in the equity REIT market. *Real Estate Economics*, 24(4), pp. 493–515.

Lin, Y.C., Cho, H. and Lee, C.L., 2019. The value-added role of sector-specific REITs in Australia. *Pacific Rim Property Research Journal*, 25(1), pp. 49–72.

Moss, A., Clare, A., Thomas, S. and Seaton, J., 2017. Can sector-specific REIT strategies outperform a diversified benchmark? *Journal of European Real Estate Research,* 10(3), pp. 366–383.

Newell, G. and Wen, H., 2006. The role of non-traditional real estate sectors in REIT portfolios. *Journal of Real Estate Portfolio Management*, 12(9), pp. 155–166.

Ntuli, M. and Akinsomi, O., 2017. An overview of the initial performance of the South African REIT market. *Journal of Real Estate*, pp. 365–388.

Ro, S. and Ziobrowski, A.J., 2011. Does focus really matter? Specialized vs. Diversified REITs. *Journal of Real Estate Finance and Economics*, 42, pp. 68–83.

15

APPLICATION OF BUILDING MANAGEMENT SYSTEM (BMS) IN COMMERCIAL PROPERTY MANAGEMENT IN KENYA

Walter Ogolla and Raphael Kieti

1. Introduction

Globally, the real estate industry has seen tremendous technological innovations in design, construction and property management. In an increasingly competitive world, building owners, designers and developers are now focused on delivering functionally efficient buildings that meet user requirements by incorporating technological innovations that increase operational efficiency, improve building performance and reduce the operational costs of buildings. Technological innovations in real estate have brought about the advantages of speed, timeliness, flexibility, increased efficiency and productivity of the real estate asset (Yahya, 2004). Among the technological advancements in property management in recent years is Building Management System (BMS) or Building Automation System (BAS). According to Boyd and Jankovic (1994), BMS is a computer-based control system installed in buildings that controls and monitors the buildings' mechanical and electrical equipment such as ventilation, lighting, power systems, fire systems and security systems. Brown (1990) defined Building Management System (BMS) as a system designed and implemented to regulate and monitor the various functions of a building's components and its associated plants. Allen and Remke (2008) stated that a building automation or management system is used to regulate various functions of a heating, ventilation and air conditioning (HVAC) system in addition to temperature and equipment scheduling. BMS is a combination of systems installed in a building to ensure ease of management and monitoring of components of buildings and facilities. Wang and Xie (2002) indicated that BMS should be adopted by property managers to ensure proper management performance of buildings, management and distribution of services as well as the provision of important management information. Wang (2010) highlights typical benefits of adopting BMS technologies to include increase in the rate of reliability of plant and services, low operational costs, rise in staff productivity and protection of people and equipment. Building Management System as a property management tool aims at ensuring efficiency by being a means for prompt relay of information on the functioning of the components of buildings thus ensuring speedy redress and action by property manager. Building Management System has successfully been embraced in developed countries while implementation

DOI: 10.1201/9780429279256-15

in developing countries is low. According to research by Charles Ogutu (2011), despite the benefits of BMS, uptake by commercial buildings in Kenya is low. Another study by Bernadette Wanjiru (2014) that assessed the effect of BMS on property value found that adoption of the system has a positive effect on the value of properties. Despite the benefits of adoption of BMS, there is no known study that has assessed the level of awareness and adoption of BMS in commercial property management in Kenya.

There is the need for property developers and building owners to embrace BMS as a property management tool in order to obtain the benefits identified by Ogutu (2011), which include reduction in operation and maintenance costs, higher rental prices, higher occupancy rates, remote and centralized supervision of buildings and energy efficiency, among others.

The present study established the level of awareness and adoption of BMS in commercial buildings and identified specific BMS concepts and technologies that have been embraced in commercial buildings in Kenya, as well as the challenges building owners and property managers face in implementation of BMS in commercial property management in Kenya.

2. Literature Review

2.1. Building Management System

A BMS system can begin with a single outstation. However, expansion can be done at any time through the addition of more outstations and connecting them using a simple network of communication. The system can be enhanced by linking the outstations to a personal computer (PC) which in turn runs the appropriate application software. A well-enhanced system acts as a management tool which allows monitoring and adjustment of all outstations from the central control point. Figure 15.1 shows arrangement of a basic BMS system.

2.2. Components of a BMS System

A BMS has both software and hardware components.

The hardware consists of *sensors*, *controllers* and *actuators*. Sensors read conditions are found all over the building and can be accessed through the following inputs: digital inputs, analogue inputs or pulse inputs. Digital inputs detect in case a device is switched on or off. Analogue inputs provide data on variables and can be used to tell different conditions within the building such as temperature, daylighting, humidity and carbon dioxide levels. Pulse inputs can provide data on the level of consumption by using devices like meters. Sensors can be used to perform different control systems and hence reduce initial and operational costs. Controllers form the major element of the BMS. For the BMS to be considered compatible, the right number of controllers has to be linked to the system. Controllers have interfaces to the sensors and actuators. Actuators perform various actions in the system and perform control functions through digital and analogue outputs. Digital outputs control devices that have two states, for example, turn equipment on or off. Analogue outputs can fine-tune devices to certain positions, such as dimming light. The sensors detect information that is later transmitted to the controller which then interprets the information to commands to be done by the actuators. Monitoring and adjustment of these actions is enabled by the connection of the user interface and the controllers; a good example is when the sensor detects changes in

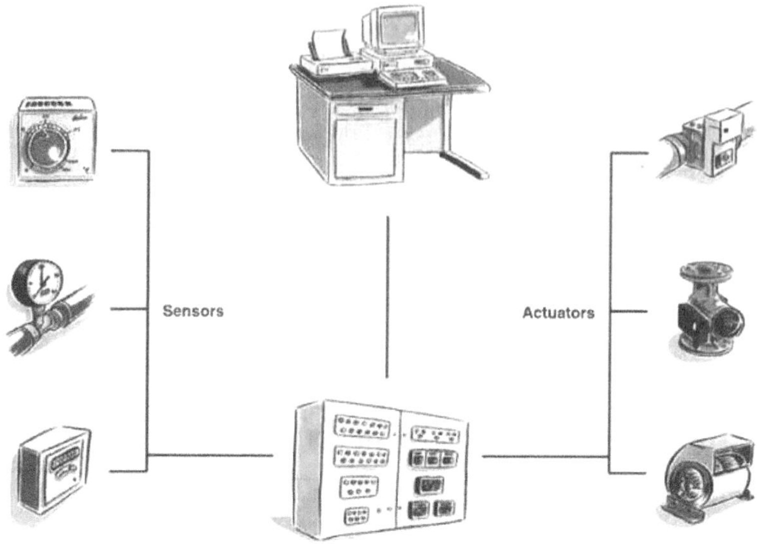

FIGURE 15.1 A basic BMS arrangement

Source: DETR Environment Transport Regions 246

the lighting levels when it is daytime, and an automatic reaction is detected through adjusting light and hence the level of energy used is reduced.

The software program is usually configured in a hierarchical manner, can be proprietary, using such protocols as C-Bus, Profibus, and so on.

2.3. Applications of BMS in Commercial Property Management

Property management is a process directed at maintaining the value of an asset or property as a resource, considering both the short- and long-term goals of the property owner (Syagga and Aligula, 1999). Property management performs an important role in the maintenance of a building by emphasizing the cost of services and condition of the building. Sound property management services play a significant role to real estate leases and sales on the market due to the fact that they guarantee both current quality of living environment and also the long-term value of a real estate asset. Effective property management ensures that a building's value is enhanced, and returns are maximized, hence property managers are on the lookout for means of enhancing the productivity of the building asset by embracing modern technology.

Building Management System (BMS) as a relatively new technology in commercial property management has been deployed in the management of commercial and other forms of real estate in both the developed and the developing countries. The technology enables property managers to monitor the property's external and internal environment through a system of coordinated sensors and actuators placed at strategic locations within the building. The output is relayed through a computer screen at the property manager's office therefore reducing the need for human to building interaction in monitoring its function. The technology has been credited for lowering building operation costs to the developer, reducing

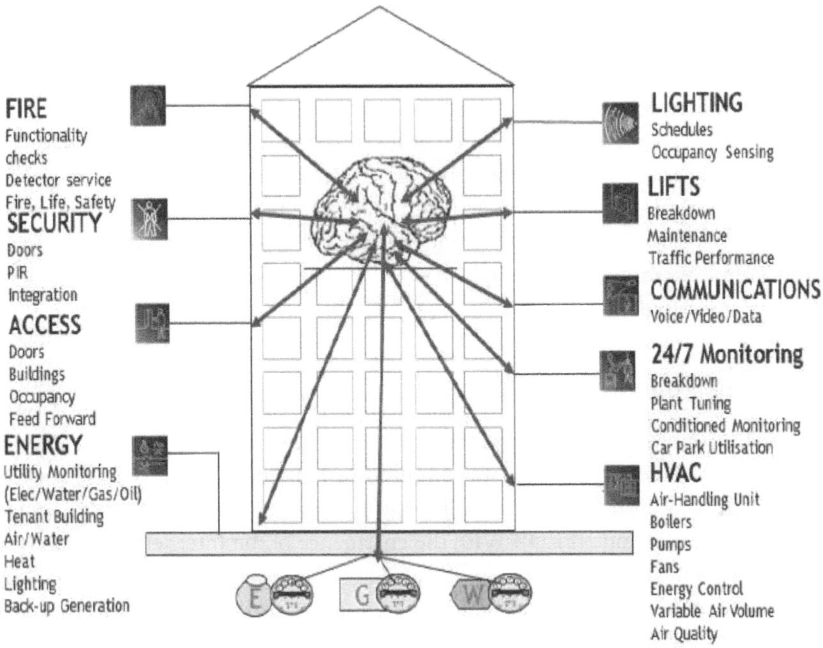

FIGURE 15.2 Applications of Building Management System (BMS) in commercial property management

Source: DETR Environment Transport Regions 246

the amount of electrical energy consumption in the building and reducing the number of personnel required for maintenance of a building.

The best practice applications of the BMS technology in commercial property management include various subsystems as discussed in the subsections that follow. All the subsystems are integrated, monitored and controlled by BMS. The applications of the subsystems are shown in Figure 15.2.

2.3.1. Fire Detection and Safety Systems

Placing fire sensors in the building enables early fire detection and increased reliability. The sensors provide data on changes that occur in the internal and external environment of a building. Such changes include humidity, temperature, movement of air, air quality, smoke, the number of individuals occupying the building as well as a host of other properties. Adopting wireless systems eliminates the need for cables and hence enables fire fighters to carry out their firefighting strategies even before arriving at the fire scene.

2.3.2. Security and Access Controls

Integrated management system (IMS) and burglar alarm subsystem forms the main elements of a building security system. These systems do not require any external support.

When set up in a building, they independently run with centralized control and can also operate efficiently with other centralized systems. Current developments and increasing terrorism threats have seen most buildings installing bio-data equipment. Users of the facilities are required to register their interests with the interests in the facility; appropriate access control smart cards are issued which ensures that only the right people have access to the specified areas of the facility (Zhang Bo, 2015). It ensures that no outsiders get into the building and interfere with activities and functions in the building. This is accomplished through the use of equipment such as finger print lock; retina based door access system; voice and video intercom; code based or swipe card access system; and biometric access system.

2.3.3. Lighting Control

Lighting is fundamental to an occupant's experience in a building. Lighting accounts for approximately 40 percent of energy consumption in buildings. The recent lighting revolution began with the push to replace traditional incandescent light bulbs with much more efficient light emitting diode (LED). With the emergence of the Internet of Things (IoT) and smart buildings came a new level of significance for lighting controls; an element now vital to achieving the potential user experience, energy saving and communicative opportunities available to the lighting sector (Saravanan and Babu Ramesh, 2016). Control methods of lighting, are as follows:

- Zones: entails switching of lights consistent to how it is used and the layout of the lit areas. This can help reduce light wastage especially when light is only needed in a small area and not the entire area.
- Time control: automatically switching on and off in every light zone to a preset schedule for light use.
- Passive Infra-Red (PIR) occupancy sensing: occupancy sensors can be applied in areas that are occupied spasmodically, to show whether somebody is in the room or not and put on or off the lights accordingly.
- Light level monitoring: this entails maintaining light level by switching or dimming artificial lighting as measured by a photocell.
- Lighting controlled motion sensors integrated with security: helps in energy reduction and increased security levels. In case the preset norms are violated, it monitors and alerts the security manager/property manager.

2.3.4. Heat Ventilation and Air Conditioning System (HVAC)

HVAC refers to how air is distributed into the workplace to ensure a comfortable environment in the building. Proper air circulation is a critical element to the operation of the building and satisfaction of the occupants. HVAC uses sensors that integrate with the automation system of the building. Through the sensors, information on the conditions throughout the building is collected. The system ensures that the levels of temperatures are conducive as well as humidity and prevents bad odor and indoor pollutants and creates comfortable air ventilation and slight temperature variation.

2.3.4. Parking Control

The parking control system consist of admission control which determines whether the vehicle is permitted to enter the parking lot and records the admission time, plate number, owner of the vehicle and other information. It also issues temporary parking cards.

2.3.5. Voice/Videocon Services

The video and voice conference aids in communication within the property and is particularly helpful for the persons with hearing and visual impairment for directions and ease of navigations. The system could also be helpful in case of emergency.

2.4. Challenges and Barriers to Adoption of Building Management System

The adoption of BMS technologies in commercial buildings can be interrupted by numerous challenges and barriers. Edwards and Grinter (2001) were the first researchers to contribute to the understanding of the key barriers preventing the adoption of smart concepts and technologies in buildings. According to the authors, the following are some of the barriers to implementation of BMS: *interoperability* means the ability of equipment, devices, appliances and systems from different vendors to operate together; *administration* refers to the expertise or experience required to maintain and operate a system; *reliability* – the interconnection of diverse technologies with different tolerances poses a risk of failure or malfunctioning; *systems intelligence and behavior inference systems* that rely on inference will never be right all of the time; and *security* – the safeguarding of personal data. As time has elapsed, other obstacles have been identified: existing homes and buildings being retrofitted, interoperability, the cost incurred during purchase and installation of smart equipment and how they are used (Holroyd et al., 2010). Additionally, Ciesielska and Li (2011) contributed to the list of barriers by adding failure to understand the needs of users and infrastructure solutions, that is, technical skills and capacity to install them as key barriers. Verbong et al. (2013) listed some institutional barriers including the domain of privacy, degree of control, lack of interest or time, and difficulty to change consumer routine behavior.

Rogers et al. (2013) grouped the barriers to implementation of BMS technologies in buildings into three categories:

- *Social barriers* that reflect a lack of knowledge about intelligent building technologies among both consumers and policy makers.
- *Financial barriers* which include the upfront cost of implementing these new technologies.
- *Structural barriers* such as incompatible communication strategies as well as availability of smart devices and the number of structural modifications required in existing buildings to implement such technologies.

In the Kenyan context, the barriers can be summarized as follows:

- Interoperability or the ability of the devices from different vendors to operate together.
- Lack of know-how on the system operations.

- Lack of reliability on the systems.
- Cost of implementation including the cost of purchase and installation.
- Lack of understanding of user needs.
- Lack of interest in the new technology.
- Lack of awareness on the new technology.
- Technical deficiencies and failures.

3. Research Design and Methodology

The research employed a case study research approach that was restricted to commercial properties located in Westlands, Upper Hill and Nairobi Central Business District (CBD) in Nairobi city in Kenya. Secondary data were obtained from review of texts, journal articles and research papers. Primary data were obtained through a census survey of property managers of five purposively selected real estate management companies in Nairobi. The choice of the companies was made considering that they are the top property management companies in the country, and they manage the largest and most recently completed commercial buildings in the study areas. A total of 59 property managers participated in the study. These were property managers employed by the management companies and actively engaged in the management of properties located in the satellite centers of Westlands, Upper Hill and Nairobi CBD. Fifty-nine questionnaires were sent to the property managers and 36 questionnaires were returned representing 61 percent response rate. Mugenda and Mugenda (2003) opine a response rate of 50 percent is adequate for analysis and reporting while 60 percent is good and 70 percent or above is considered very good. Therefore, an overall response rate of 61 percent is adequate for analysis and reporting. A great deal of information was obtained through site visits and observation of commercial buildings to ascertain the type and level of adoption of BMS technologies. Data were analyzed using descriptive statistics, and a 5-point Likert scale was used to analyze the challenges of adoption of BMS technologies. All the data analysis procedures were performed using Microsoft excel 2013, and results presented in tables, charts and figures.

4. Data Analysis and Results

The study examined the level of awareness and adoption of Building Management System (BMS) concepts and technologies in commercial properties in Kenya. The study also aimed at identifying specific BMS technologies implemented in commercial properties and determining challenges and barriers in the implementation of these technologies, as well as strategies that could be used to ensure increased level of adoption of BMS technologies in Kenya.

The study results are presented under the headings, as follows:

4.1. Level of Awareness and Adoption of Building Management System (BMS) Concepts and Technologies

One of the objectives of the study was to examine the level of awareness and adoption of BMS concepts and technologies. The technologies were grouped into seven categories namely, heating, ventilation and air conditioning (HVAC); lighting controls; security and observation controls; electrical power controls; Elevators control; fire and alarm system; and plumbing and water monitoring. The property managers were asked if they understood the concept of centralized BMS and whether they are familiar with the BMS concepts and technologies. The results are as tabulated in Table 15.1 and Figure 15.3.

The results show that only 28 percent of the respondents are familiar and understood the concept of centralized BMS and its role in commercial property management. The level of awareness of BMS concepts and technologies was varied with lighting control having the highest level of awareness at 83 percent of the respondents, while security and observation controls was second with 78 percent level of awareness. When the respondents were requested to state the commercial buildings in their portfolio that had fully adopted a Building Management System, all the management companies had no building with a fully functioning BMS. Overall, the study findings revealed a low level of awareness of centralized BMS among property managers at 28 percent but an average level of awareness of MBS concepts and technologies at 64 percent.

4.2. BMS Technologies and Concepts Implemented in Commercial Buildings in Kenya

The study sought to find out specific BMS technologies and concepts that have been implemented in commercial buildings in Kenya. Property managers were asked to indicate the BMS technologies that had been implemented in the buildings they were managing. The results are presented in Table 15.2 and Figure 15.4.

The Table 15.2 and Figure 15.4 indicate the percentage of adoption of BMS concepts and technologies in commercial buildings in Kenya. The results reveal that security and observation controls, and fire and alarm system were fully implemented in the studied buildings both having 100 percent level of implementation, while elevator controls and electrical power control were third and fourth respectively at 75 percent and 61 percent implementation levels. None of the surveyed buildings had installed a heating, ventilation and air conditioning (HVAC) system in the building; installation of air conditioning alone was, however, reported but only installed where the tenant had requested. This can be attributed to the moderate climatic conditions experienced within the study areas in Nairobi.

TABLE 15.1 Awareness of BMS and BMS Concepts and Technologies

Building Management System Concepts and Technologies	*Awareness of the BMS Concepts/Technologies*		*% Awareness*
	Yes	*No*	
Centralized Building Management System (CBMS)	10	26	28
Heating, Ventilation and Air Conditioning (HVAC)	20	16	56
Lighting Control	30	6	83
Security and Observation Controls	28	8	78
Electrical Power Control	26	13	72
Fire and Alarm System	25	11	69
Elevators	27	9	75
Plumbing and Water Monitoring	18	18	50
Total Average Level of Awareness			**64**

Source: Field survey, 2019

as stated by Ciesielska and Li (2011). Lack of documented local experience scored third as a hindrance to the adoption of BMS with a mean score of 3.93 while consideration of such technologies as a luxury by developers ranked fourth with a mean of 3.69 and lack of understanding of user needs ranked fifth with a mean of 3.67. Respondents were of the view that studies needed to be done to isolate critical user needs for technologies in commercial buildings. Technical deficiencies including lack of support staff in case of breakdown of the technologies ranked sixth as a hindrance to adoption of BMS technologies with a mean of 3.58 while lack of awareness of BMS benefits and lack of knowhow of the systems operations ranked seventh and eighth with means of 3.53 and 3.11, respectively.

4.4. *Strategies to Facilitate Widespread Adoption of BMS*

The study further asked the respondents to give their opinions on what should be done to ensure widespread adoption of BMS. Respondents were expected to rate the opinions on a scale of 1 to 5 with 1 being the factor that would lead to least widespread adoption and 5 being the factor that would lead to highest rate of adoption; the ratings were ranked by calculation of mean score of stated opinions and ranked. The results are displayed in Table 15.4 and Figure 15.6.

The study findings indicated that certain initiatives should be put in place to enable widespread adoption of BMS in commercial properties in Kenya. Tax waivers on BMS technologies ranked highest with a mean of 4.33; the respondents noted that tax waivers would lower the cost of importing this system and thus make them readily available. Adoption of regulations enforcing adoption of BMS technologies ranked second with a mean of 3.92. Respondents had the opinion that laws enforcing adoption would speed up the rate of BMS adoption. Increased public awareness on the need and benefits of BMS and introduction of training on BMS at institutions of higher institutions ranked third and fourth with means of 3.36 and 3.31, respectively. Financial incentives to developers who have implemented BMS ranked fifth with a mean of 3.22. The respondents stated that property developers considered financial incentives should be on tax laws and not direct financial refunds.

5. Conclusion and Recommendations

The main objective of the study was to examine the level of awareness and adoption of Building Management System, concepts and technologies in commercial properties in Kenya. The study also aimed at identifying specific BMS technologies implemented in

TABLE 15.4 Strategies to Facilitate Widespread Adoption of BMS

Strategies to Facilitate Adoption of BMS	*Mean*
Tax waivers on Building Management System technologies	4.33
Laws guiding and regulating adoption of BMS technologies	3.92
Public awareness on the need and benefits of BMS	3.36
Introduction of BMS courses in institutions of higher learning	3.31
Financial incentives to developers who have implemented BMS	3.22

Source: Field survey, 2019

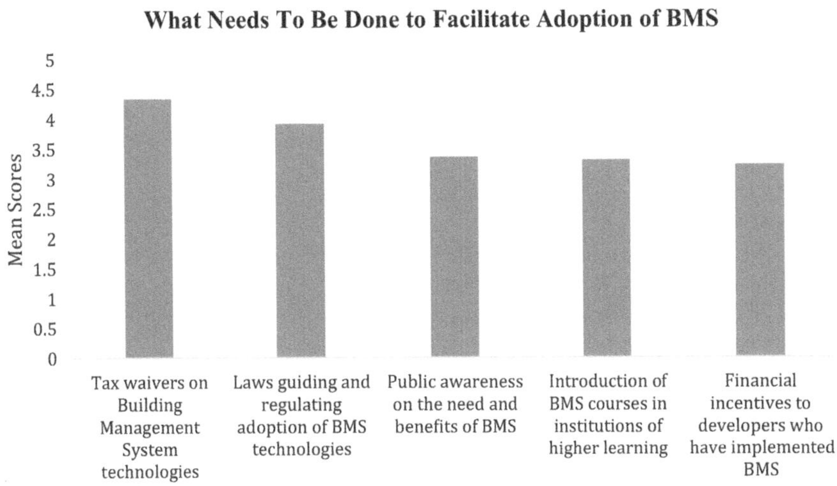

FIGURE 15.6 Strategies to facilitate adoption of BMS

Source: Field survey, 2019

commercial buildings and challenges in the adoption of the technologies, as well as strategies for increased adoption of the technologies. The objectives of the study have been achieved. The study results revealed a low level of awareness and adoption of centralized BMS and an average awareness of specific BMS concepts and technologies among property managers. Security and observation controls, and fire and alarm system have widely been adopted and implemented in commercial buildings, while heating, ventilation and air conditioning (HVAC) is the least adopted BMS technology in commercial buildings. Lack of availability of low cost BMS technologies, high purchase and installation costs, lack of documented local experience among others, are major deterrents to adoption and implementation of BMS technologies in Kenya. The study recommends that the government should introduce measures such as tax waivers on BMS technologies and enactment and enforcement of laws guiding and regulating adoption of BMS technologies, as strategies for increased adoption and implementation of BMS concepts and technologies in commercial properties in Kenya.

References

Allen, P. J. and Remke, R. (2008). Building automation systems (BAS): Direct digital control. *Encyclopedia of Energy Engineering and Technology*, Vol. 2, pp. 104–110.

Boyd, D. and Jankovic, L. (1994). Building IQ: Rating the intelligent building. In Boyd, D. (ed.) *Intelligent Buildings*, Alfred Waller Limited in association with Unicom, London.

Brown, R. G. (1990). Building management system. *Property Management*, Vol. 8, No. 3, pp. 212–219. Emerald Publishing Ltd, UK.

Ciesielska, M. and Li, F. (2011). The connected home: From market barriers to business model solutions. In Skersys, T., Butleris, R., Nemuraite, L. and Suomi, R. (eds.) *Building the e-World Ecosystem. I3E 2011. IFIP Advances in Information and Communication Technology*, Vol. 353. Springer, Berlin, Heidelberg. https://doi.org/10.1007/978-3-642-27260-8_15.

Edwards, W. K. and Grinter, R. E. (2001). *At Home with Ubiquitous Computing: Seven Challenges*. Computer Science Laboratory, Xerox Palo Alto Research Center Palo Alto, California 94304.
Holroyd, P., Watten, P. and Newbury, P. (2010). Why is my home not smart? Aging friendly technology for health and independence. In Lee, Y., Bien, Z. Z., Mokhtari, M., Kim, J. T., Park, M., Kim, J., Lee, H. and Khalil, I. (eds.) *Aging Friendly Technology for Health and Independence*, Vol. 6159, pp. 53–59, Springer-Verlag, Berlin.
Mugenda, O. M. and Mugenda, A. G. (2003). *Research Methods: Quantitative and Qualitative Approaches*. Acts Press, Nairobi.
Ogutu, C. O. (2011). The status of intelligent buildings in Kenya case study – commercial facilities. Unpublished Dissertation, University of Nairobi, Kenya.
Rogers, E., Elliott, R., Kwatra, S., Trombley, D. and Nadadur, V. (2013b). *Intelligent Efficiency: Opportunities, Barriers, and Solutions*. ACEEE, Washington, DC. Available at aceee.org/research-report/e13j.
Saravanan, S. and Babu Ramesh, N. (2016). Maximum power point tracking algorithms for photovoltaic system- a review. *Renewable and Sustainable Energy Reviews*, Vol. 57, pp. 192–204, Pergamon.
Syagga, P. M. and Aligula, E. M. (1999). *Management and Agency in the Urban Real Property Market in Kenya*. Berichteaus der Volkswirtschatf, Germany.
Verbong, G. P., Beemsterboer, S. and Sengers, F. (2013). Smart grids or smart users? Involving users in developing a low carbon electricity economy. *Energy Policy*, Vol. 52, pp. 117–125.
Wanjiru, B. K. (2014). Concepts of building automation and effect of the property value in Kenya. Unpublished Dissertation, University of Nairobi, Kenya.
Wang, S. (2010). *Intelligent Buildings and Building Automation*. Spon Press, London.
Wang, S. and Xie, J. (2002). Integrating building management system and facilities management on the Internet. *Automation in Construction*, Vol. 11, No. 6, pp. 707–715. ISSN 0926-5805, Department of Building Services Engineering, The Hong Kong Polytechnic University, Kowloon, Hong Kong, China.
Yahya, A. S. (2004). The application of information technology in commercial property management in Nairobi. Unpublished Master's Thesis, University of Nairobi.
Zhang Bo. (2015). Intelligent building security system design based on internet of things technology. 4th National Conference on Electrical, Electronics and Computer Engineering (NCEECE).

16

AN INVESTIGATION INTO THE PERCEIVED IMPACT OF IMPLEMENTING INNOVATIVE DIGITAL TECHNOLOGY IN THE PROPERTY TRANSACTION PROCESS IN SOUTH AFRICA

Implications for African Property Markets

Jason Green, François Viruly and Alireza Moghayedi

Blockchain	Fully distributed cryptographic system which consists of a digital ledger of transactions which have taken place between network actors and are stored on a consistent, immutable and linear chain. (Karamitsos *et al.*, 2018).
Innovative Digital Technology	Emerging advanced digital technology such as, *inter alia*, blockchain, smart contracts and artificial intelligence.
Property Transaction Process	The end-to-end process of buying or selling property including, *inter alia*, marketing, conveyancing, payment and transfer of ownership.
Smart Contracts	Digital programmable contracts which are enforced once a predefined set of conditions have been successfully executed (Nowiński and Kozma, 2017).

Acronyms:

DLT	Digital Ledger Technology
IDT	Innovative Digital Technology
PTP	Property Transaction Process

1. Introduction

An immense amount of capital remains inaccessible and slow moving in South Africa and across the African continent where the majority of the population live in rural areas, often without formal ownership of property or with a lack of access to affordable property related services (Medina *et al.*, 2017). This lack of service, specifically with respect to property transactions, is typically the result of the low value of property in this bracket and the associated low financial reward anticipated by property transaction intermediaries in exchange for

DOI: 10.1201/9780429279256-16

facilitating the execution of these transactions (Mooya and Cloete, 2007). High property transaction costs prevent the exchange of property in these instances, which results in market failure (Mooya and Cloete, 2007). The negative effects of the inefficiencies of the process are not only experienced by low-value property owners. Higher value property owners in a position to transact property must endure a process which can require more than 100 transactions between fragmented intermediaries to be executed before the transaction is successful, a process which typically takes between two and six months to complete (Amadi-Echendum and Pellissier, 2014).

The relative transparency of African property markets is illustrated in the JLL Global Real Estate transparency index (JLL and LaSalle, 2020). It shows that South Africa is the only transparent property market across the African property market – ranked 24th of the 99 countries surveyed. Kenya and Mauritius fall in the semi-transparent category with other countries across the continent falling in the low and opaque categories.

The World Bank ease of doing business report (World Bank, 2020) illustrates that in sub-Saharan Africa some 6.1 procedures need to be followed to register a property which can be compared to 4.7 procedures in OECD high income countries. Moreover, while it takes on average 23.6 days to transfer properties in OECD high income countries, it takes 51.6 to transfer a property in sub-Saharan Africa.

The number of tech hubs across the African continent has increased from 314 in 2016 to 442 in 2018. In total a record US$725.6 million was raised across 458 projects in 2018, representing a massive 300% increase over 2017. These include accelerators, incubators and co- working spaces (URERU, 2019). There is growing acceptance that new technologies can effectively allow African economies to leapfrog more industrialized countries in the application of technologies.

The emergence of the Fourth Industrial Revolution (4IR) and the associated exponential rate of development of novel innovative digital technology (IDT) around the globe provides an opportunity to address the inefficiencies of property transaction processes. As such, this research attempts to answer the following question: What is the perceived impact that the implementation of innovative digital technology, such as blockchain and smart contracts, could have on the property transaction process in South Africa? The aim is to investigate the perceived impact that the implementation of innovative digital technology could have on the inefficiencies of the property transaction process in South Africa, along with the identification of advantages, disadvantages and barriers to the implementation of this technology.

2. Literature Review

2.1. Technological Development

Humankind is currently at the forefront of the 4IR as digital technologies become more integrated and sophisticated, fueling transformation in both industry and society (Schwab, 2016). Further, the emergence of the 4IR and associated technology such as blockchain, artificial intelligence (AI), Internet of Things (IoT) and machine learning is identified by Baum *et al.* (2020) as the drivers which could enable PropTech 3.0. These technologies underpin the business change evident in the PropTech sector. Different characteristics have been used to define the PropTech environment. Baum *et al.* (2020) characterize the sector as the digital disruption of the industry, and the application of digital data. Related business models include the collection, aggregation and processing of information, which in turn affects the use, management and ownership of property. Two specific technologies investigated for the purposes of this research are blockchain or distributed ledger technology, and smart contracts. The key characteristics of blockchain are summarized by Zheng *et al.* (2018) as decentralization, persistency, anonymity and auditability. It can be considered that blockchain is a wider application of Digital Ledger Technology (DLT), incorporating the storage of information along with the facilitation of transactions, while DLT is narrower in its application. Smart contracts are effectively computer programs which make decisions based on the inputs received and certain pre-programmed conditions being met (Kolvart *et al.*, 2016). Kolvart *et al.* (2016) note that the main advantage of smart contracts is their potential impact on efficiency of processes which result from the self-enforceable nature of these contracts.

The broad array of role players involved in the property transaction results in a complex network, particularly when taking into account the lack of integration of internal processes and technological systems between role players (Amadi-Echendu and Pellissier, 2013). Further, Graglia and Mellon (2018) note that "entire industries have emerged for the sole purpose of capitalizing on the inefficiencies of the property transfers." It is argued that this process is constrained by the manual interlinks and information transfer loops which have been established between the various role players (Amadi-Echendum and Pellissier, 2014).

The illiquid nature of real estate leads to bottlenecks in the property transaction process. Outside of broader traditional processing bottlenecks, technological constraints such as internal database systems and servers used by the various role players also experience bottlenecking issues in the processing high transaction volumes (Liu *et al.*, 2020). Baum *et al.* (2020) describe the property asset class to be slow moving as a whole. H. de Soto (2000), cited by Toulmin (2009), describes the long and bureaucratic nature of sub-Saharan property transactions in particular, and their tendency to slow transaction times and support corruption. Gumede and Dipholo (2014) argue that public administration is "very slow and suffers from lack of greater responsiveness, allocative inefficiency, distributive incapability and weak re-distributive efficacy in South Africa." These in turn increase transaction costs and associated risks. With respect to transparency, Baum *et al.* (2020) note its importance in achieving the efficient transfer of property ownership as a means of providing role players access to the information necessary to make rational, informed decisions. Baum *et al.* (2020) go on to attribute the inefficiency of property transactions to the prevalent information asymmetry resulting from a lack of transparency. From a cost perspective, the cost of participating in a formal market is broadly broken up into three categories – search and information costs, bargaining and decision costs, and supervision and enforcement costs (Furubotn and Richter, 2010). These transaction costs accumulate over the course of a property transaction, and if the gains from the transaction do not sufficiently outweigh the costs of the transaction, there is the potential for market failure to occur (Mooya and Cloete, 2007). Financial institutions have established a number of processes in an effort to address the risk of fraud. Amadi-Echendum and Pellissier (2014) describe these processes as having significant time (and therefore cost) implications. Furthermore, Boles (2017) describes the prevalence of opportunities for the laundering of money in the South African real estate market through the purchase or development of luxury real estate with cash.

2.2. *Implementing Innovative Digital Technology in the Property Transaction Process Through Blockchain Technology*

Levin (2018) describes the overcoming of fragmentation in the public sector as one of the key compelling arguments of the convergence of technologies which characterize the 4IR in South Africa. The decentralization characteristic of blockchain technology lends itself to the defragmentation of processes. Furthermore, transactions conducted on blockchain networks can be conducted between peers without interaction with a central agency (Zheng *et al.*, 2018). This is supported by Swan (2015) who identified the potential for blockchain technology to disintermediate and decentralize transactions processes of all types involving all parties. Baum and Saull (2019) note that intermediaries such as conveyancers and property managers are most at risk in terms of having their roles disrupted through the implementation of IDT in property transactions.

Nowiński and Kozma (2017) argue that the decentralized ledger technology which forms the basis of blockchain technology reduces the costs of administration and provides a mechanism to reducing the processing time of transactions. In addition, the digital transparency achieved by notarizing and publicly recording transactions via hashing timestamps can significantly increase the difficulty of committing fraud due to the increased difficulties in falsifying records (Graglia and Mellon, 2018). Further, the ability to leverage user verification tools on a blockchain system allows transparency to be addressed which can lead to improved accountability of stakeholders and ultimately the improved speed of transaction processing (Nowiński and Kozma, 2017). Kshetri (2017) highlights the benefit that blockchain transaction could have on high transaction costs where the transaction is effectively conducted by the transacting parties themselves without the reliance on third parties and without the associated fees. With respect to security, the digitalization of the property transaction process will require specific responsiveness to the risks associated with data security, privacy and cyber security (Levin, 2018). In this respect, Kshetri (2017) describes blockchain technologies' first direct benefit as its potential to reduce corruption and fraud. Furthermore, Kolvart *et al.* (2016) note that one of the key characteristics of smart contracts is their ability to mitigate the risks associated with illegal activity and accidental error.

Kumar *et al.* (2020) describe blockchain technology as a "high-cost and high-overhead storage medium" and associate it with high initial set up costs, due to the early stage of development and learning curve associated with the technology, and high transaction or operational cost due to validation and consensus mechanisms. In a developing nation context, and with respect to the low levels of education and lack of skills necessary to use emerging technology including blockchain, Batubara *et al.* (2018) identify the disparity evident in comparison to developed nations. This is supported by Ayentimi and Burgess (2019) who note the emerging skill shortages and inadequate investment in skills development and education. From a legislative standpoint, Amadi-Echendu and Pellissier (2013) identify the need for relevant statutes to be amended in order to cater for the use of digital signatures in South Africa. Another major legal barrier to the implementation of blockchain in real estate transactions identified by Baum and Saull (2019) relates to the "legal recognition of blockchain-based contracts." Levin (2018) describes the need for an enabling environment in order for industries to adopt technologies at scale. The need to create a robust legal framework to regulate areas impacted by the use of IDT is a government-led action which will ultimately determine whether the adoption of this technology can successfully occur at scale (Leurent and de Boer, 2018).

2.3. The African Context

The World Bank (2020) describes a rise in poverty in South Africa in recent years, with unemployment having reached an 11-year high of 29.1% in the third quarter of 2019 and 19.8% of the population living in extreme poverty (below US$1.90 per day). Poverty and unemployment are directly linked to economic growth, and Wong *et al.* (2005) argue that the driving impact of technological innovation on economic growth has long been established based on the theoretical contributions of Solow (1956) and Romer (1986), along with the empirical evidence described by Mansfield (1972) and Nadiri (1993). Gross Domestic Expenditure on Research and Development (GERD) as a percentage of Gross Domestic Product (GDP) provides a useful measurement of innovation on a national scale. According to the NACI (2019) the latest measure of GERD as a percentage of GDP was 0.82% as at 2016/2017. In comparison, NACI (2019) notes the 1.4% spent by upper middle-income countries (UMIC) in 2015 and the 1.5% global GERD as a percentage of GDP in the same year. These figures illustrate the fact that South Africa is lagging behind UMIC's with respect to investment in research and development at the impairment of innovation and ultimately economic growth. To further inform the state of innovation in the African context, GERD in low income and middle-low income countries was 0.29% and 0.43% in 2017 respectively (NACI, 2020). A further indication of this disparity is evident in the fact that South Africa accounts for 77.4% of scientific publications in the Southern African Development Community (SADC) region (NACI, 2020).

The Centre for Housing Finance (CAHF) reports that the South African government has built in excess of 3 million RDP houses, of which a mere 1.9 million houses have been registered. The national Department of Human Settlements, Water and Sanitation (NDHSWS) estimates that the title deed backlog for RDP properties built prior to 2014 stands at 511 752 (Africa News, 2019).

A study which looked at the South African PropTech environment (URERU, 2019) identified 42 active players in the PropTech sector. These players were categorized into the following segments identified by Baum *et al.* (2020), namely crowdfunding and finance, coworking, digital innovation, selling and letting, services and software, and data analytics. The number of players per category is illustrated in Table 16.1:

TABLE 16.1 South Africa PropTech segmentation

Sector	*Number of players*
Crowd Funding and Finance	2
Co-working	4
Digital Innovation	6
Selling and Letting	13
Services and Software	12
Data Analytics	5

Source: URERU, 2019. Many of these entities attempt to address market inefficiencies and the reduction of transaction costs.

The acceptance of innovative digital technologies has been slow in South Africa, with much of the focus having been placed on the marketing of residential properties. Innovative technologies have been accepted by South African commercial banks who have adopted automated valuation systems. Seso Global, a blockchain property registry company, is introducing a blockchain-based property registry (Africa News, 2019). This relatively slow uptake of innovative technologies has *inter alia* been influenced by the size of the local market as well as the number of players in the market.

3. Research Methodology

This research is based on primary data collected by means of semi-structured interviews administered to property and technology experts. Five interviewees participated in this research. These participants were identified and selected based on the following criteria: experience in the South African property market; experience in property transactions; and awareness of emerging innovative digital technology. Semi-structured interviews were selected as the primary means of collecting data for two reasons. First, they allowed for "the exploration of the perceptions and opinions of respondents regarding complex and sometimes sensitive issues and enabled probing for more information and clarification of answers" (Barriball and While, 1994). Second, the broad nature of the topic being investigated and the wide range of stakeholders involved required a flexible interviewing method in order to accommodate their varied backgrounds, expertise and experiences. The data collected were then analyzed by means of thematic analysis in line with the six step, three stage framework proposed by Attride-Stirling (2001). Furthermore, secondary data which was collected and explored in the literature review was then contrasted against the analysis of the data collected from the interviews from which conclusions were derived.

4. Findings and Discussion

4.1. South African Property Transaction Process Inefficiencies

The fragmentation of the property transaction process (PTP), which is largely attributed to the lack of structural links and integration of systems between role players, was identified as a key inefficiency by the interviewees. This is supported by Amadi-Echendu and Pellissier (2013), Bell *et al.* (2018) and Liu *et al.* (2020) who all refer to the fragmentation of the PTP as a driver of inefficiency. Processing time emerged as another key inefficiency in the process, where this was largely attributed to the fact that it is a generally analogue process. This was supported by the literature with Gumede and Dipholo (2014) and Amadi-Echendum and Pellissier (2014) highlighting the role of the Deeds office in contributing to the slow processing of property transfers. Furthermore, the manual information transfer loops identified by Amadi-Echendum and Pellissier (2014) contribute further to the slow processing.

Data collection, management and analysis emerged as another key inefficiency. Supporting views were identified in the literature with Amadi-Echendu and Pellissier (2013) in particular raising concern regarding the centralized and fragmented storage of data. In addition, Saull *et al.* (2020) and Amadi-Echendum and Pellissier (2014) highlight the processes established by institutions to address the risk of fraud.

Moreover, the analogue nature of the process with the potential for human error also emerged as another inefficiency of the PTP. This is supported by Amadi-Echendu and Pellissier (2013) and Mooya and Cloete (2007) who described the manual nature of the existing process which results from the current systems employed and the susceptibility of these systems to becoming complicit to the inefficiencies experienced.

High transaction costs with respect to the legal conveyancing profession and the public sector emerged as a key inefficiency in the process. This is in alignment with the literature, where Mooya and Cloete (2007) identified different types of transaction costs which accumulate over the course of a property transaction and largely included legal and enforcement fees.

The disparity between formal and informal markets with respect to having their property needs met emerged as a final inefficiency by the interviewees. This was largely associated with the lack of linkage to the formal PTP due to the inefficiency of the process in catering for the informal market's needs. SERI (2018) highlighted the lack of service in informal markets. This is supported by Mooya and Cloete (2007) with respect to market failure occurring if the costs of the process outweigh the benefits for low income households.

4.2. Perceived Impact of Innovative Digital Technology on South African Property Transaction Process Inefficiencies

The potential to streamline the process through the integration of intermediaries' systems and the disintermediation of the process through the application of innovative digital technology (IDT) emerged in the interviews. This is supported by the literature with Levin (2018) identifying the defragmentation of public sector participants as a compelling argument. Further, Zheng *et al.* (2018) highlighted that transactions conducted on a blockchain framework enable peer-to-peer interaction without the need for third party participation.

The potential to reduce the processing time of the PTP through the application of IDT was identified by the research participants. This perception is supported by the literature where Amadi-Echendu and Pellissier (2013) identified the potential to link intermediaries to a central system in order to improve the speed of transactions, while Nowiński and Kozma (2017) advocate for IDT such as DLT and smart contracts in overcoming process administrative burdens and speeding up the PTP.

The potential for IDT to improve the transparency and accountability of the PTP emerged. The transparency which could be achieved was also associated with accountability which could be facilitated through the application of IDT. This is supported in the literature by Nowiński and Kozma (2017); Graglia and Mellon (2018) and Amadi-Echendu and Pellissier (2013) with respect to making information more readily available and driving accountability in the PTP through the application of IDT.

It emerged that the documentation process could be smoothed significantly through the application of IDT and the subsequent elimination of error by reducing the number of hands which touch the data could be realized. The potential to improve this inefficiency through the application of IDT is highlighted by Amadi-Echendu and Pellissier (2013).

The potential to improve the collection, management and usage of data involved in the PTP was highlighted as a potential impact. The implementation of a blockchain system was identified as having the potential to create a verifiable and immutable bank of data. Further, IDT's potential to address the inefficient nature of the authentication and vetting processes emerged. This is supported by the literature with Amadi-Echendu and Pellissier (2013) and Spielman

(2016) who argue that the application of IDT can address the authentication and vetting inefficiencies of the PTP by leveraging private keys as personal identification when interacting with the blockchain property transaction network, acting as a single source of identification.

A reduction in transaction costs, as a result of the improved processing, disintermediation and improved information flow emerged as a potential impact. This is aligned with the literature, where Kshetri (2017) and Graglia and Mellon (2018) note the benefit that blockchain technology would have in reducing transaction costs by reducing the reliance on third parties and their associated fees. In addition, Amadi-Echendu and Pellissier (2013) note that IDT could reduce transaction costs associated with the storage of hardcopy documentation, costs linked to lost or misfiled title deeds or costs linked to the cancellation of bonds due to the reliance on a digital system.

4.3. Advantages of Implementation of Innovative Digital Technology

Security of transactions, through the application of a secure blockchain and smart contract framework emerged as a potential advantage by the interviewees. This is supported in the literature by Kshetri (2017), Levin (2018), Mehdi (2020) and Kolvart *et al.* (2016) who note the positive impact that the implementation of IDT will have on security, fraud prevention and the mitigation of risks associated with illegal activity and accidental error. The implementation of IDT could also result in the development of a new sector within the property industry. In addition, from management and advisory perspective, more opportunities will arise as a result of increased demand due to the low levels of familiarity with IDT. Finally, the potential to improve security of tenure was identified where it was held that more stability in the country could be achieved through the decentralization of the PTP and the resulting reduction of government interference.

4.4. Disadvantages of Implementation of Innovative Digital Technology

The potential for job losses emerged as a potential consequence of the implementation of IDT. This was largely attributed to the redundancies that could materialize through the disintermediation of both public and private role players. In addition, the research participants held that the internet may not be the most secure environment in which to transact. In this sense, cyber security and the potential corruption of the system by external forces emerged as a potential disadvantage. This is supported in the literature by Mehdi (2020) who identified system architecture of the blockchain network as what will ultimately define its vulnerability. In addition, weak passwords, malware or the interference of bad network actors were identified as potential risks to the security of the system, and therefore the security of transactions. Further, the potential for the implementation of IDT in PTP to drive further inequality emerged. The knowledge barrier, with respect to working with new technologies, coupled with limitations of accessibility to technology and connectivity may drive further inequality.

4.5. Barriers to the Implementation of Innovative Digital Technology

The capital costs involved in the initial development and implementation of IDT in the PTP was identified as a major barrier by the research participants. The high capital costs associated with not only the technology, but the skills development aspect contribute to the deterrence perceived by both the public and private sector. As such, a lack of incentive was identified from both a public and private sector perspective. This is supported by the literature, where Leurent and de Boer (2018) argue that incentives which support the development of this technology are a critical consideration which will drive the adoption of IDT. Furthermore, Kumar *et al.* (2020) identify the high capital and initial set up costs linked to the early stage of development of blockchain technology, coupled with the steep learning curve associated with the technology.

Technology operating costs also emerged as a barrier to the implementation of IDT. Effectively, the highly skilled nature of the personnel who would need to manage and operate the IDT may increase operating costs. In addition, from a data storage efficiency standpoint, blockchain was identified as a relatively expensive means to store information. This view was supported by the literature, where Kumar *et al.* (2020) identify the high overhead costs and high operational costs associated with the validation and consensus mechanisms prescribed by blockchain technology.

The perceived lack of education, expertise and skills from a technological standpoint in South Africa was identified as another major barrier. Low digital literacy rates, skills shortages and a sharp learning curve were attributed to this perception. In addition, the need for cross-industry skills requirements, such as IT and accounting, further exasperate the significance of this barrier. This is aligned with the findings in the literature, where Batubara *et al.* (2018) and Ayentimi and Burgess (2019) highlighted concerns that skills shortages and low levels of investment in education will pose challenges to the implementation of IDT and the emergence of the 4IR.

The internal focused nature of role players in the PTP with respect to protecting their income streams and intellectual property or data emerged as another barrier to the implementation of IDT in the PTP. This is largely due to the fact that any integration of services will require the sharing of data and cooperation between role players. This was supported in the literature through Baum *et al.* (2020) who noted that transaction intermediaries may be hesitant to adopt modern approaches due to the negative impact it could have on their bottom line. The lack of suitable legislation to facilitate the implementation of IDT in the PTP, along with a lack of government support was identified by the participants as major barriers. From a legislative perspective, it was noted that there is no legislation, with the exception of the Protection of Personal Information Act 4 of 2013, to support the implementation. Government's lack of political will and comprehension of the potential of IDT also contribute to the barrier perceived. This view is supported by the literature, where Leurent and de Boer (2018), Baum and Saull (2019) and Levin (2018) describe the need for an enabling environment from a legislative point of view, and that a robust legal framework is required in order to drive the legal recognition of blockchain-based smart contracts.

Connectivity, with respect to access to internet, the availability of bandwidth and access and affordability of data, was identified as a barrier to the potential implementation of IDT in the PTP. Further, effective change management, which would be required for such a transformative process, was highlighted as a barrier in the sense that it has the potential to prevent the successful implementation of IDT in the PTP. Finally, the availability of technology was identified as a barrier to the implementation of IDT largely due to the fact that the IDT in question is in its infancy in terms of development.

5. Conclusion and Further Research

This research has established the perceived positive impact that the implementation of IDT could have on the inefficiencies of the South African PTP. The key inefficiencies identified were the fragmentation of the PTP; slow processing time; data collection and management challenges; the authentication and vetting process; the analogue nature of the process and associated risk of human error; high transaction costs; and the informal market. Further, the perceived impact that IDT could have on these inefficiencies included the defragmentation and disintermediation of the PTP; improved processing time; improved transparency; improved documentation processing; improved data collection and management; improved authentication and vetting processes; lower transaction costs; and the formalization of informal markets.

In addition, auxiliary advantages and disadvantages emerged. The perceived advantages identified were the reduction of fraud and illegal activity; new job opportunities and job creation; and improved security of tenure. The disadvantages identified included job losses and skillset redundancies; susceptibility to cyber-crime; and increased inequality and class divide. Furthermore, barriers to the implementation of IDT in the transaction process emerged, namely high capital and initial costs associated with the technology; high operating costs associated with the management and operation of the technology; a shortage of the necessary skills, knowledge and expertise; the agency issue and internal focused nature of existing transaction role players; the role of government and lack of suitable legislature; availability of the technology; connectivity and infrastructure restraints; and the need for effective change management. The aforementioned outcomes of this research will contribute to the existing gap in the literature regarding IDT and its implementation in property transactions in a South African context.

Areas identified for further research which emerged over the course of this research study include: the development of a roadmap illustrating the technical implementation of IDT in the PTP in South Africa should be developed; the development of a framework explicitly identifying points of inefficiency within the existing process should be developed; quantitative research should be undertaken in order to quantify the inefficiency of the existing transaction process and measure the improvement possible through the implementation of innovative digital technology; further research into the feasibility and viability of the proposed implementation of technology in a South African context should be undertaken; and an investigation into the existing legislative impacting to the implementation of IDT should be undertaken, with the intention of developing a robust legal framework.

While this study provided an insight into the implementation of innovative digital technology in South Africa, it also suggests the opportunities and risks that might exist in implementing similar technologies across the African continent. The introduction of innovative digital technologies has the potential to improve market transparency through a reduction of transaction costs, and by improving the 'bankability' of properties. An improvement in market transparency also makes it easier for domestics and international investors to consider the role of African properties in diversified portfolios.

References

Africa News (2019) South Africa's first ever blockchain-based property register pilot [Online]. Available www.africanews.com/2019/11/06/south-africas-first-ever-blockchain-based-property-register-pilot/. Accessed 05 May 2021.

Amadi-Echendu, A. P. and Pellissier, R. (2013) Lessons for South Africa from the international e-conveyancing environment. In *Global Business and Technology Association's 15th Annual International Conference*, Helsinki, Finland, July 2–6, pp. 2–6.

Amadi-Echendum, A. P. and Pellissier, R. (2014) Eliminating bottlenecks in the South African conveyancing environment. *Mediterranean Journal of Social Sciences*, **5**(14), 97–103.

Attride-Stirling, J. (2001) Thematic networks: An analytic tool for qualitative research. *Qualitative Research*, **1**(3), 385–405.

Ayentimi, D. T. and Burgess, J. (2019) Is the fourth industrial revolution relevant to sub-Sahara Africa? *Technology Analysis & Strategic Management*, **31**(6), 641–652.

Bank, T. W. (2020) *Macro Poverty Outlook: South Africa*. The World Bank.

Barriball, K. L. and While, A. (1994) Collecting data using a semi-structured interview: A discussion paper. *Journal of Advanced Nursing-Institutional Subscription*, **19**(2), 328–335.

Batubara, F. R., Ubacht, J. and Janssen, M. (2018) Challenges of blockchain technology adoption for e-government: A systematic literature review. In *Proceedings of the 19th Annual International Conference on Digital Government Research: Governance in the Data Age*. Association for Computing Machinery, New York, NY, USA, Article 76, pp. 1–9.

Baum, A. and Saull, A. (2019) *The Future of Real Estate Transactions*. Saïd Business School, University of Oxford.

Baum, A., Saull, A. and Braesemann, F. (2020) *PropTech 2020: The Future of Real Estate*. Saïd Business School, University of Oxford.

Bell, J., Goga, S., Mondliwa, P. and Roberts, S. (2018) *Structural Transformation in South Africa: Moving Towards a Smart, Open Economy for All*. CCRED Working Paper No. 9/2018 [Online]. Available https://ssrn.com/abstract=3269732 or http://dx.doi.org/10.2139/ssrn.3269732

Boles, J. R. (2017) Anti-money laundering initiatives for the South African real estate market. *JCULP*, **1**, 197.

Furubotn, E. G. and Richter, R. (2010) *Institutions and Economic Theory: The Contribution of the New Institutional Economics*. University of Michigan Press.

de Soto, H. (2000) *The Mystery of Capital: Why Capitalism Triumphs in the West and Fails Everywhere Else*. Basic Books.

Graglia, J. M. and Mellon, C. (2018) Blockchain and property in 2018: At the end of the beginning. *Innovations: Technology, Governance, Globalization*, **12**(1–2), 90–116.

Gumede, N. and Dipholo, K. (2014) Governance, restructuring and the new public management reform: South African perspectives. *Journal of Educational and Social Research*, **4**(6).

JLL and LaSalle. (2020). *Global Real Estate Transparency Index, 2020*. JONES LANG LASALLE IP, INC.

Kolvart, M., Poola, M. and Rull, A. (2016) Smart contracts. In *The Future of Law and Etechnologies*. Springer, pp. 133–147.

Kshetri, N. (2017) Will blockchain emerge as a tool to break the poverty chain in the global south? *Third World Quarterly*, **38**(8), 1710–1732.

Kumar, A., Liu, R. and Shan, Z. (2020) Is blockchain a silver bullet for supply chain management? Technical challenges and research opportunities. *Decision Sciences*, **51**(1), 8–37.

Leurent, H. and de Boer, E. (2018) *The Next Economic Growth Engine: Scaling Fourth Industrial Revolution Technologies in Production*. World Economic Forum.

Levin, S. (2018) *World Economic Forum and the Fourth Industrial Revolution in South Africa*. Trade & Industrial Policy Strategy, TIPS.

Liu, N., Duncan, D. and Chapman, A. (2020) *A Critical Review of Distributed Ledger Technology and Its Applications in Real Estate*. RICS Research Trust.

Medina, L., Jonelis, M. A. W. and Cangul, M. (2017). *The Informal Economy in Sub-Saharan Africa: Size and Determinants*. International Monetary Fund.

Mansfield, E. (1972) Contribution of R&D to economic growth in the United States. *Science*, **175**(4021), 477–486.

Mehdi, N. (2020) *Blockchain: An Emerging Opportunity for Surveyors?* RICS.

Mooya, M. M. and Cloete, C. E. (2007) Informal urban property markets and poverty alleviation: A conceptual framework. *Urban Studies*, **44**(1), 147–165.
NACI (2019) *South African Science, Technology and Innovation Indicators*. NACI.
NACI (2020) *South African Science, Technology and Innovation Indicators*. NACI.
Nadiri, M. I. (1993) *Innovations and Technological Spillovers*. 0898–2937. National Bureau of Economic Research.
Nowiński, W. and Kozma, M. (2017) How can blockchain technology disrupt the existing business models? *Entrepreneurial Business and Economics Review*, **5**(3).
Romer, P. M. (1986) Increasing returns and long-run growth. *Journal of Political Economy*, **94**(5), 1002–1037.
Saull, A., Baum, A. and Braesemann, F. (2020) Can digital technologies speed up real estate transactions? *Journal of Property Investment and Finance*, **38**(4), 349–361. https://doi.org/10.1108/JPIF-09-2019-0131
Schwab, K. (2016) *The Fourth Industrial Revolution*. Portfolio Penguin.
SERI (2018) *Informal Settlements and Human Rights in South Africa*. Socio-Economic Rights Institute of South Africa.
Solow, R. M. (1956) A contribution to the theory of economic growth. *The Quarterly Journal of Economics*, **70**(1), 65–94.
Spielman, A. (2016) *Blockchain: Digitally Rebuilding the Real Estate Industry*. Massachusetts Institute of Technology.
Swan, M. (2015) *Blockchain: Blueprint for a New Economy*. O'Reilly Media, Inc.
Toulmin, C. (2009) Securing land and property rights in sub-Saharan Africa: The role of local institutions. *Land Use Policy*, **26**(1), 10–19.
URERU (2019) *Mapping the East African Proptech Ecosystem*. URERU.
Wong, P. K., Ho, Y. P. and Autio, E. (2005) Entrepreneurship, innovation and economic growth: Evidence from GEM data. *Small Business Economics*, **24**(3), 335–350.
Zheng, Z., Xie, S., Dai, H.-N., Chen, X. and Wang, H. (2018) Blockchain challenges and opportunities: A survey. *International Journal of Web and Grid Services*, **14**(4), 352–375.

17

REAL ESTATE EDUCATION IN AFRICA SOUTH OF THE SAHARA

Catherine Kariuki, Kathy Michell and Rachael Daisy Mirembe

1. Introduction

Real estate education has, at its core, certain universal and value-neutral skills and knowledge. This knowledge has developed over decades to reflect the functions of specific market structures which characterise formal real estate markets. It is generally acknowledged that dominant real estate theory and related practices are based on mature markets within an industrialised context and do not accurately reflect real estate markets and decision making in real estate markets in emerging economies (Mirembe & Viruly, 2018). Kampamba et al. (2015) argued that standardisation of real estate education is not possible due to differences attributed to specific industry needs, different tenure systems and backgrounds of those developing a curriculum. The important issue is that curriculum should attempt to reflect the market context in which it is delivered. While countries that are characterised by mature real estate markets tend to place a strong emphasis on course units such as investment and finance, this should not be the case for countries characterised by more opaque markets (emerging markets) due to specific industry needs, and different land tenure systems. There is a need for African researchers to merge western acquired knowledge, skills, methodologies and tools of research with African reality (Nsamenang, 1995 in Ezeanya-Esiobu). Indigenous knowledge is critical for African advancement in all fields and sectors, real estate included (Ezeanya-Esiobu, 2019). It would be good for academic and other stakeholders in the real estate industry to first assess the needs in their respective countries before designing new or reviewing their curricula (Cloete, 2002).

The real estate industry in Africa is characterised by a low level of market maturity (Dugeri, 2011; Rothacher, 2013; Olaleye & Adebara, 2019). Immature markets present different challenges. For Instance, in valuation of properties which is one of the cores of the real estate curricula, the assets to be valued are often unusual and have limited institutional infrastructure. All these additional risks of unusual assets, unknown covenants and high levels of volatility may not be adequately reflected in values. There is also the case of unregistered land. Over 90% of land in Africa is not registered and is primarily held under customary rights. Consequently, the valuation methods that are taught and practiced in the western countries,

DOI: 10.1201/9780429279256-17

for example the United Kingdom where most of their land is registered, may not be wholly relevant when applied in Africa. Moreover, it means the financing options for real estate that are used in the western countries may not apply in Africa because the collateral security is questionable. Therefore, the investment option is also affected. Subsequently, it can be argued that the land and property rights issue alone in Africa determines the nature and functioning of the real estate industry in Africa and therefore governs the curriculum content of the universities that teach real estate.

This chapter is a continuation of the previous study by Mirembe & Viruly, 2018 where eight universities were assessed to establish the kind of curriculums that are used. In that study, the duo found that real estate programmes in sub-Saharan Africa (SSA) tend to find an appropriate balance between subjects of a social and financial nature. Yet the research also suggested that it is difficult to identify strong similarities between real estate programmes offered in SSA. In this study, our major aim is to identify the similarities and differences that exist in 17 real estate programmes across Africa, and then to assess the gaps that need to be filled up in relation to the African context. To this end, the following section reviews the literature about real estate education and the different approaches that have been adopted in the different parts of the world. This is followed by a section on the methodology that was utilised for the study. The fourth section is the discussion and lastly, conclusions are drawn, and recommendations made with respect to how real estate education should be taught in Africa.

2. Literature Review

Over the years, there have been calls to standardise the real estate curricula but due to the changing nature of real estate, this is has not been achieved (Kampamba et al., 2015). As a result, it has become increasingly difficult to define the body of knowledge of real estate education (Black & Rabianski, 2003). There appears to be no clear cut consensus on the boundary lines of the discipline (Black & Rabianski, 2003; Jayantha & Chiang, 2012). Often the analysis arises from specific philosophical perspectives: American, UK or the multi-disciplinary. For instance the American approach places a strong emphasis on finance with a vast majority of real estate academics being housed or affiliated with business colleges (Roulac, 2002). The American approach underlines that if one is to understand real estate, then scholars ought to be exposed to multiple disciplines of business (Roulac, 2002). On the other hand, the UK approach mainly focuses on the classic surveying curriculum with strong emphasis on land related disciplines. Typical courses include valuation, law, economics, building construction, planning, management and information technology (Dasso & Woodward, 1980; Schulte & Schulte-Daxbök, 2000; Galuppo & Worzala, 2004). The UK approach is better at exposing students for hands on involvement with land and buildings placing lesser emphasis on seeing real estate purely as an investment class.

In Brazil, India, China and Australia, real estate courses have been organised in a diffused and heterogeneous manner with varying emphasis (Schulte & Schulte-Daxbök, 2000). Some of the major themes include real estate investments, financial systems, economic evaluation, project management, appraisal, land economics and information technology. Thus, real estate courses in these countries can be regarded as adopting a multi-disciplinary approach. Some scholars have supported the need for the multi-disciplinary approach as it brings out both the advantages of the American and UK approach (Black et al., 1996; Graeme et al., 2004; Musil, 2005; Mooya, 2007).

Moreover, due to a lack of a common body of knowledge, its usual to see programmes of the annual real estate conferences with research on widely diverse topics and real estate schools with degree programmes consisting of very different subject matter or different paradigms and body of knowledge (BOK) (Dasso & Nourse, 1992; Epley, 1996). Epley 2006 elaborates on the different existing curriculum paradigms that are being used to teach real estate programmes in universities as shown in Table 17.1.

The existence of the different paradigms is evidence that real estate area has not evolved through the detailed debate that is required to delineate the topics and receive consensus

TABLE 17.1 Existing curriculum paradigms

Curriculum Paradigm	*Comment*
Four-course textbook model	• Follows traditional four- academic course approach for an undergraduate degree. • Involves adopting a text, and teaching whatever is in the book. This typically includes a course in real estate principles, finance, law and appraisal.
Two- to four-course model	• Same as the preceding one, emphasis in real estate except a major does not exist.
Elective area within the MBA	• The MBA is structured to allow electives, one to four courses in real estate can be offered.
Specialisation with an MS in real estate	• A Master of Science in real estate consists of 24–30 graduate hours in real estate topics only.
Terminal degree in real estate	• Consists of a wide variety of concentrations. • The degrees include a Ph.D. in real estate that is awarded by a real estate department, Ph.D. in another field such as finance with a major in real estate, Ph.D. in economics with a concentration in urban and regional economics, Ph.D. in architecture, and a DBA in finance with a major in finance.
Emphasis on development	• Emphasis and teaching on real estate development. • Urban Land Institute has supported this approach.
Emphasis on land economics	• Historically has been the basis upon which the subject of real estate was built on for four-year university courses. • The Appraisal Institute educational material and BOK can be viewed as being based in land economics. • Vestiges of this subject are still found in courses in regional economics and selected chapters in the principles and appraisal courses.
Emphasis on analysis and decision making	• Since the practice of real estate in today's market is transaction-based, concentration is on analysis and decision making. • Recommendations have been made to include additional courses like asset management and corporate real estate analysis and management.
Emphasis on a systems or management approach	• Concentration on management approaches of the various subspecialties like financing the transaction, estimating its value and employing an attorney to assure that the asset is in the marketplace at its highest and best use
Emphasis on an employer need approach	• Suggests that the BOK be designed according to the needs of our current employers. • The employers' opinions should be solicited to determine the content and emphasis of the curriculum.
Emphasis on AACSB requirements	• This model suggests that the overriding concern for real estate acceptability is approval by the AACSB for one or more courses.

Source: Epley, 1996, Pg. 232–233

among educators about the necessary knowledge and skills required (Epley, 1996). There have been some studies to investigate successful elements of real estate programmes. Carn and Rabianski (1986) recognised five core subjects which should form part of a real estate curriculum such as investment analysis, real estate finance, market analysis and appraisal, policy, and law. Weeks and Finch (2003) investigated several undergraduate core curricula in universities and found that real estate programmes in colleges of business had a focus on economic and financial constructs. Ford and Elkes (2008) suggest that real estate principles and real estate finance are the most important for undergraduates in real estate. Other topics that have been suggested to be included in the real estate curriculum include, but are not limited to, information technology, entrepreneurship, corporate strategy, mortgage securitisation, urban form, technology advances, environmental concern, globalisation, strategic resource and public-private concern. All these studies have been carried out in the developed countries like the USA and UK. These topics suggested are only reflective of the maturity of real estate markets in the developed world which are transparent, with high level of capital liquidity and many real estate products being traded (Keogh & D'Arcy, 1994). These topics do not accurately reflect real estate markets in Africa. The generic context of real estate markets in Africa tends to be characterised by a dual economy, a lack of transparency in the markets, high levels of uncertainty and the existence of communal/customary and informal rights in the ownership of real estate. Since the aim of the curriculum is to create effective real estate decision makers and managers who are equipped with concepts, techniques and skills that are vital for problem solving in today's world and in the future, then real estate theories and taught curricula should reflect these characteristics and the level of maturity of the markets they attempt to analyse, as well as the institutional market arrangements, political and economic environment. The problems that affect Africa in the real estate industry are different from those in the developed countries. The way universities prepare students for real estate in Africa must be consistent and reflect the way graduates will work and solve real estate problems on the African continent. For example, in many developing countries it has been noted that 90% of land rights are not registered. Yet land is a key resource in accessing financial services that will ultimately lead to economic growth and poverty reduction. Valuation of this unregistered land still needs to be undertaken because of the large land acquisitions being done for infrastructure and large-scale investments in food and biofuels (UN-Habitat, 2018). Academic programmes have been left behind in the preparing approaches for valuation of unregistered land. They have left it to UN-Habitat, the World Bank, RICS. These organisations have set global goals to address these issues under the Sustainable Development Goals. Furthermore, in the case where universities have attempted to change the curriculum to give students an opportunity of specialisation in valuation, property management, land and housing administration (Nzioki et al., 2006) the students felt it was reducing their choices of employment. The specialisation had been seen as a means to increasing the utility of the course. This was rejected and the specialisations removed.

3. Methodology

Previous scholars who have studied real estate curriculums largely carried out their studies using surveys and internet searches (Butler et al., 1998; Black & Rabianski, 2003; Galuppo & Worzala, 2004; Schulte et al., 2005; Epley, 2006; Chikafalimani et al., 2012; Jayantha & Chiang, 2012; Kampamba et al., 2017). These projects relied on surveys distributed to professionals and academics. This was often complemented with secondary data from educational institutions that teach real estate studies. It also included the verification of data from university websites. Table 17.2 shows the existing studies on real estate curriculum and the different methodologies used to achieve the study aims.

TABLE 17.2 Summary of existing studies on real estate curriculum and methodologies used

Authors, Year	*Title of Study*	*Methodology Used*
Butler et al., 1998	Integrating the real estate curriculum	Questionnaire
Schulte, 2001	Real estate education at the European business school	Document analysis
Schulte, 2002	Real estate education throughout the world: past, present, and future	Internet searches and focus groups
Black & Rabianski, 2003	Defining the real estate body of knowledge: a survey approach	Questionnaires
Galuppo & Worzala, 2004	A study into the important elements of a master's degree in real estate	Literature review, focus groups and questionnaires
Schulte et al., 2005	Internationalisation of real estate education	Literature reviews, scanning of websites and questionnaires
Epley, 1996	The current body of knowledge paradigms used in real estate education and issues in need of further research	Questionnaires
Nzioki et al., 2006	Education and training in valuation, land management and housing in Kenya	Record searches and analysis
Mc cuddy et al., 2008	Using student feedback in designing student focused curricula	Questionnaire
Poon et al., 2011	Real estate education: an investigation of multiple stakeholders	Interviews and questionnaires
Chikafalimani et al., 2012	Analysis and comparison of master of real estate curricula in South Africa	Internet searches, record searches and analysis
Jayantha and Chiang, 201	Key elements of successful graduate real estate education in Hong Kong: students' perspective	Questionnaire survey
Poon, 2014a	Commercial awareness in real estate courses	Interview and questionnaire
Poon, 2014b	Do real estate courses sufficiently develop graduate's employability skills?	Interview and questionnaire
Kampamba et al., 2015	A comparative analysis of real estate education in Botswana	Internet searches, interviews, record search and analysis, questionnaire
Kampamba et al., 2017	An evaluation of the relevance of real estate curricula in Botswana	Questionnaires
Amidu et al., 2018	Exploring gaps between real estate curriculum and industry needs: a mapping exercise	Desk top review

In this study, university websites were checked for information concerning their relevant programmes to assess the differences that may exist between the different programmes. After that, real estate academics in different universities in Africa were contacted to provide the researchers with the necessary programme backgrounds. Seventeen academics responded and were involved in the survey. The 17 universities include University of Cape Town; University of the Witwatersrand, in South Africa; Copperbelt University in Zambia; Namibia University of Science and Technology; University of Malawi; University of Nairobi; Technical University of Kenya; Kenyatta University; Mount Kenya University; Ardhi University in Tanzania; Makerere University in Uganda; Federal University of Technology in Minna; Obafemi Awolowo University, Federal University of Technology, Akure, University of Lagos, all in Nigeria; and Kwame Nkrumah University of Science and Technology in Ghana. An analysis of the 17 universities was conducted and the university courses in real estate were then categorised into eight subcomponents based on literature (Mirembe & Viruly, 2018). The eight categories include: Business in Real Estate; Law and Taxation; Construction and Technology; Business and Management; Finance and Valuation; Economics; Land Surveying and Environment; and Others. The different courses under each of the eight categories are shown in Table 17.3.

4. Discussion of the Findings

4.1 Summary of the Categorisation of Real Estate Programmes in Africa

The programmes offered in most of the universities demand a high level of theoretical engagement and intellectual independence and aim to foster development of deepened, comprehensive and systematic expertise in the major cognate areas of learning. Students are equipped with cognitive and intellectual skills, key transferable skills and professional, technical, practical skills that would enable them to promote and manage properties within working, living, recreational and commercial contexts. The programmes focus on a comprehensive and analytical understanding of property valuation methods, facilities and property management, property marketing, property economics and finance and capability of making sound and well-informed judgments in line with relevant legislation, thus the grouping or categorisation of the course units into the following categories: Business in Real Estate, Finance and Valuation, Law and Taxation, Economics, Construction and Technology, Land Surveying and Environment, Business and Management and others as tabulated in Table 17.4.

Business in Real Estate course units' categories included facilities management, property maintenance, housing, real estate principles, real estate market analysis, real estate management, real estate development, brokerage business management, development perspective, property development, estate management, corporate real estate management, real estate marketing and agency and real estate information system. The university with the highest percentage of these courses is the University of the Witwatersrand and University of Lagos at 23%.

For the Finance and Valuation category, the following course units were undertaken: mathematics for real estate, valuation, accounting, financial management, investment appraisal analysis, land taxation, value and risk management, real estate finance, business finance, management accounting, risk and insurance practice, and corporate finance. These units seem

TABLE 17.3 Categorisation of real estate programmes in sub-Saharan Africa

Business in RE	*Finance and Valuation*	*Law and Taxation*	*Economics*	*Construction/ Technology*	*Land Surveying and Environment*	*Business and Management*	*Others*
• Facilities Management. • Property Maintenance. • Housing. • RE Principles. • RE Market Analysis. • RE Management. • RE Development. • Brokerage Business Management. • Development Perspective. • Property Development. • Estate Management. • Corporate RE Management. • RE Marketing and Agency. • RE Information System.	• Mathematics for R.E. • Valuation. • Accounting. • Financial Management. • Investment Appraisal and Analysis. • Land Taxation. • Value and Risk Management. • RE Finance. • Business Finance. • Management Accounting. • Risk and Insurance Practice. • Corporate Finance.	• Elements of Law and Government. • Law of Contract and Tort. • Law of Real Property. • Business Law. • Property Dispute Resolution. • Commercial Law. • Real Estate Law. • Labour Law. • Property and Contract Law. • Land Law. • Planning and Public Policy Analysis. • Landlord Tenant Law.	• Micro Economics. • Macro Economics. • Agriculture and Economic • Management Land Economics. • Natural Resource Economics. • Housing Economics and Administration. • Econometrics for Property Studies. • Urban Economics. • Property Studies.	• Architectural Drawing and Design. • Building Technology. • Science and Technology in Development. • Building Materials and Finishes. • Project Management. • Measurement. • Architectural Studies and Construction. • Construction. • Building Materials. • Infrastructure and Building Services. • Management Structural and Condition Surveys. • Built Environment. • Free Hand Sketching.	• Physical Environment. • Environmental Science. • Environmental Building Services. • Introduction to Land Surveying. • Elements of Urban and Regional Planning. • Land Administration and Information Systems. • Environmental Impact Assessment and Audits. • Land Management and Policy Studies. • Planning for Property Developers. • Urban Land Use and Development, • Geographical Information Systems. • Land Information Systems.	• Communication Skills. • Introduction to Computing. • Information Computerised Technology. • Principles of Management. • Entrepreneurial Studies. • Contracts & Procurement in Real Estate Services. • Professional Ethics. • Management and Leadership. • Evidence Based Management. • Marketing. • Strategic • Management. • Human Resource Management. • Value Management.	• Statistics. • Mathematics. • Quantitative Methods. • English Language. • Research Methods. • Science and Technology in Development. • Principles of Agriculture and Forestry. • Field Attachment. • Globalisation and the Built Environment. • Industrial Training. • Philosophy. • HIV Aids.

Source: Mirembe, R., & Viruly, F. (2018). Real Estate Knowledge and the Development of a Real Estate Curricula for African Universities. Page 5, *Journal of African Real Estate Research*, 3(2), 9–17.

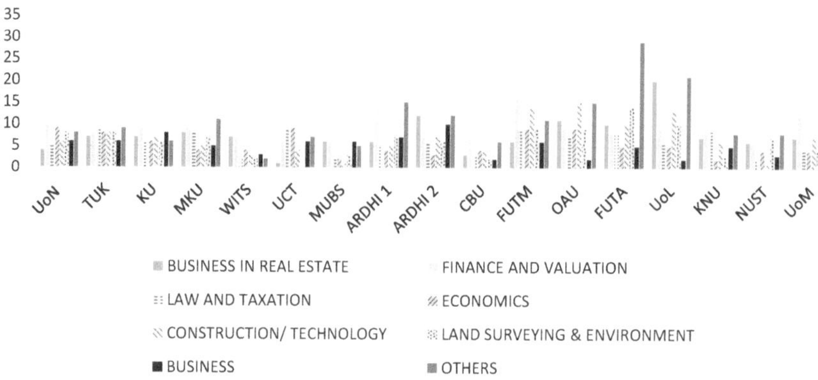

FIGURE 17.1 Categorisation of Real Estate Programmes in Sub-Saharan Africa

to have the highest percentage in most of the universities with the highest being 23% for University of the Witwatersrand and Copperbelt University, 22% for Ardhi and 20% to 21% for Kwame Nkrumah University of Science and Technology and University of Malawi respectively. At least five universities are between 16% and 19%, while the balance five are between 9% and 14%.

The Law and Taxation category included the following course units: elements of law and government, law of contract and tort, law of real property, business law, property dispute resolution, commercial law, real estate law, labour law, property and contract law, land law, planning and public policy analysis, and landlord and tenant law. The university with the highest number of units in this category is University of Cape Town at 21%, followed by Kwame Nkrumah University of Science and Technology at 18% with the lowest being University of the Witwatersrand, Makerere University and Namibia University of Science and Technology at 7%, 6% and 5% respectively.

In the Economics category, the following course units were included: micro economics, macro economics, agriculture and economic, management and land economics, natural resource economics, housing economics and administration, econometrics for property studies, urban economics and property studies. University of Cape Town has the highest number in Economics at 21% followed by University of Nairobi at 16%.

The Construction and Technology category included the following course units: architectural drawing and design, building technology, science and technology in development, building materials and finishes, project management, measurement, architectural studies and construction, construction, building materials, infrastructure and building services, management structural and condition surveys, built environment and free hand sketching. These ranged from 3% to 19%. The University of Obafemi Awolowo University having the highest followed by Federal University of Technology in Estate Management at 18%. While the University of Lagos is at 15%. These universities offer a degree in estate management.

The Land Surveying and Environment category included the following course units: physical environment, environmental science, environmental building services, introduction to land surveying, elements of urban and regional planning, land administration information systems, environmental impact assessments and audits, land management and policy studies, planning for property developers, urban land use and development, geographical information

TABLE 17.4 Percentages of subject categorisation per univeristy

S/No.	University	Programme	Categories of Real Estate Courses																	
			Business in Real Estate		Finance and Valuation		Law and Taxation		Economics		Construction and Technology		Land Surveying and Environment		Business and Management		Others		Total Number of Courses	
			No.	%	No.	%	No.	%	No.	%	No.	%	No.	%	No.	%	No.	%	No.	%
1	**University of Nairobi**	**Bachelor of real estate**	4	7%	10	18%	5	9%	9	16%	6	11%	8	14%	6	11%	8	14%	56	100%
2	**Technical University of Kenya**	**Bachelor of real estate**	7	11%	8	13%	9	14%	8	13%	8	13%	8	13%	6	10%	9	14%	63	100%
3	**Kenyatta University**	**Bachelor of Science in real estate**	7	13%	9	16%	6	11%	6	11%	7	13%	6	11%	8	15%	6	11%	55	100%
4	**Mount Kenya University**	**Bachelor of Science in real estate management**	8	14%	9	16%	8	14%	4	7%	5	9%	7	12%	5	9%	11	19%	57	100%
5	**WITS**	**Bachelor of Science in property studies**	7	23%	7	23%	2	7%	4	13%	3	10%	2	7%	3	10%	2	7%	30	100%
6	**UCT**	**Bachelor of Science in property studies**	1	2%	6	14%	9	21%	9	21%	4	10%	0	0%	6	14%	7	17%	42	100%
7	**Makerere University**	**Bachelor of real estate business management**	6	19%	6	19%	2	6%	2	6%	1	3%	3	10%	6	19%	5	16%	31	100%
8	**Ardhi University**	**Bachelor of Science in real estate finance and investment**	6	10%	14	22%	5	8%	4	6%	5	8%	7	11%	7	11%	15	24%	63	100%
9	**Ardhi University**	**Bachelor of Science in property and facilities management**	12	19%	7	11%	6	10%	3	5%	7	11%	6	10%	10	16%	12	19%	63	100%
10	**Copper Belt University**	**Bachelor of Science in real estate studies**	3	10%	7	23%	3	10%	4	13%	4	13%	2	6%	2	6%	6	19%	31	100%

(Continued)

TABLE 17.4 (Continued)

S/No.	*University*	*Programme*	*Categories of Real Estate Courses*																	
			Business in Real Estate		*Finance and Valuation*		*Law and Taxation*		*Economics*		*Construction and Technology*		*Land Surveying and Environment*		*Business and Management*		*Others*		*Total Number of Courses*	
			No.	*%*	*No.*	*%*	*No.*	*%*	*No.*	*%*	*No.*	*%*	*No.*	*%*	*No.*	*%*	*No.*	*%*	*No.*	*%*
11	**Federal University of Technology – Mina**	**Bachelor of Technology in estate management**	6	8%	16	20%	9	11%	9	11%	14	18%	9	11%	6	8%	11	14%	80	100%
12	**Obafemi Awolowo University**	**Bachelor of Science in estate management**	11	14%	9	12%	7	9%	9	12%	15	19%	9	12%	2	3%	15	19%	77	100%
13	**Federal University of Technology – Akure**	**Bachelor of Technology in estate management**	10	11%	8	9%	8	9%	5	6%	10	11%	14	16%	5	6%	29	33%	89	100%
14	**University of Lagos**	**Bachelor of Science in estate management**	20	23%	9	10%	6	7%	5	6%	13	15%	10	12%	2	2%	21	24%	86	100%
15	Kwame Nkrumah University	**Bachelor of Science in estate management**	7	14%	10	20%	9	18%	2	4%	6	12%	3	6%	5	10%	8	16%	50	100%
16	**Namibia University of Science and Technology**	**Bachelor of property studies**	6	16%	6	16%	2	5%	4	11%	1	3%	7	19%	3	8%	8	22%	37	100%
17	**University of Malawi**	**Bachelor of Science in land economy – honours**	7	12%	12	21%	4	7%	4	7%	7	12%	4	7%	4	7%	15	26%	57	100%

systems and land information systems. Namibia University of Science and Technology leads in this category at 19%, followed by Federal University of Technology, Akure at 16% with University of Nairobi at 14%.

In the Business Management category, the following course units included communication skills, introduction to computing, information computerised technology, principles of management, entrepreneurial studies, contracts and procurement in real estate services, professional ethics, management and leadership, evidence-based management, marketing, strategic management, human resource management and value management. Makerere University has the highest number in this category at 19%, followed by Kenyatta University at 15%, then University of Cape Town at 14%.

The Other course unit's category known as Common units include communication skills, introduction to computer applications, development studies, HIV/AIDS and drug abuse, entrepreneurship, statistics, mathematics, quantitative methods, English language, research methods, science and technology in development, principle of agriculture and forestry, field attachment, globalisation and the built environment, industrial training and philosophy.

Despite the Federal University of Technology – Akure offering 89 units, 16 of these units are elective. Some of the courses offered addressed specific country concerns including Elements of Law I and II. These address the customary landholding system in Nigeria, modes of acquiring title to land, settlement, Kola tenancy[1] among others. The rating and taxation unit looks at the history and evolution of rating in Nigeria, local government and its functions and land use change in Lagos. The land administration course aims at introducing students to land administration problems and subsequent transactions in formal and informal land markets. How the land in the informal markets is valued is not evident in the Applied Valuation Unit II. We can only assume that such land will only be valued under the New Valuation Approaches which are not specified.

Upon completion of the programmes, graduates of these programmes find employment in the real property – marketing, agency, real property valuation, assessment of value and risk, investment appraisal, facilities and estate management. Others join the academic and research fields.

The graduates are expected to be able to:

1. Locate relevant property management and value information sources and perform various analytical functions such as financial and market analysis and interpretation relating to real estate.
2. Produce basic building drawings for the purposes of appraising and assessing land and buildings values.
3. Plan, value different types of properties for various purposes using several methods and produce valuation reports in response to a specific "client" need as per recognised standards.
4. Demonstrate deepened understanding of time value of money and its applications to capitalisation and discounting in the evaluation of alternative property investments.
5. Evaluate economic issues pertaining to the property sector at both micro and macro level.
6. Appreciate the role of the property development process and the property finance options available.
7. Apply ethics in the context of codes of professional conduct and legal liability.

8. Oversee the management, maintenance and repair work on residential properties, shopping centres, office complexes and other facilities.
9. Perform the role of a mediator during the sale and rental of various types of properties and provision of consultancy services.
10. Demonstrate the generic graduate outcomes of problem-solving, critical thinking, responsible citizenship and good communication.
11. Plan and conduct supervised research of an applied nature in the field of real estate.
12. Communicate key principals and concepts relevant to a wide array of issues pertinent to real estate.
13. Apply knowledge, skills and experiences acquired to undertake valuations of moveable and immoveable assets, feasibility studies, investment appraisals and risks assessments.
14. Formulate practical solutions in estate management, facilities management, and property marketing and agency.
15. Demonstrate professionalism in engaging with other stakeholders in the profession.
16. Analyse research findings and innovative solutions to address the myriad and complex problems arising from compact real estate property developments.

From the analysis, most of the universities have adopted either the American approach which is more bent towards business or the UK approach which is more land related or even the interdisciplinary approach which mixes both the American and UK approaches. The only few universities that take cognisance of African issues in their curricula are the Kwame Nkrumah University of Science and Technology and University of Lagos. Kwame Nkrumah University of Science and Technology in Kumasi, Ghana introduces two subjects that clearly take cognisance of emerging economies. The course in principles of customary law that is taught is aimed at introducing the student to the customary rules and regulations which govern land tenure relations in Ghana. In being able to identify the general rules of customary land tenure, the student recognises the differences among various tribal communities in Ghana. Another course is the literature in English, which has two units. The objective of the first unit is to introduce the student to opportunities to a love of books in English and various Ghanaian languages. The second unit in literature includes folktales, short stories and the novel.

The degree in estate management and valuation of the University of Lagos responds directly to local situations including The National Taxation and Rating I and II that introduces the student to the history of rating and taxation in Nigeria. The course unit in comparative land policies looks at the Nigerian Tradition Land Policies and a unit in mineral valuation. This unit introduces the students to mineral deposits in Nigeria, taking special cognisance that Nigeria is an oil producing country.

5. Conclusion and Recommendations

Since the aim of the curriculum is to create effective real estate decision makers and managers who are equipped with concepts, techniques and skills that are vital for problem solving in today's world and in the future, then real estate theories and taught curricula should reflect these characteristics and the level of maturity of the markets they attempt to analyse, as well as the institutional market arrangements, political and economic environment. The problems that affect Africa in the real estate industry are different from those in the developed countries. The way universities prepare students for real estate in Africa must be consistent and reflect the way graduates will work and solve real estate problems on the African continent. For instance, there are still challenges such as communal/customary and informal rights. There is still a lot of land that is unregistered or owned communally. As we come up with new curriculum for real estate or review the existing ones, we ought to think of issues such as equipping our students with techniques and skills that are vital for solving our African problems. There is a need to think of the institutional market arrangements, political, social and economic environment while considering the level of maturity of the African market. It does not benefit a student in Uganda to have a whole property finance course concentrating mainly on real estate investment trusts (REITS) as an investment option when the country's stock exchange does not have even a single real estate investment item listed on it. Instead have property finance with REITS as one of the topics but concentrate on other topics such as how to raise funds in an environment where the collateral security is nonexistent since the land is communal. That would mean that in the early years of the real estate programme, the curriculum should have introduced the student to the customary land tenure system in depth and maybe the laws that exist to govern such property rights. This kind of thinking has not been seen widely in the 17 curriculums that were examined and analysed. As we come up with the real estate curricula, we ought to understand the institutional context in which our property markets in Africa operate. Unlike the western countries where formal rules exist and are implemented, in Africa, most of our property markets are not regulated. This simply means that informal rules are more dominant in these markets compared to the formal rules. As a result, the way real estate business is carried out in such markets is different. So, our curricula should focus on helping our students understand the informal rules in our markets, who the players in the market are, and how you become a professional real estate agent/broker in a market where we have "roadside brokers" and "briefcase" real estate agents. How do you survive in a market as a professional where there is free entry and exit from the real estate industry, where there are no professional bodies to regulate the behaviour of the professionals, where the government bodies are not helping in the smooth running of the real estate industry?

If the curriculum can include such issues and many more that are not discussed in this book chapter but are happening in African real estate markets, then we will be able to produce real estate graduates with the relevant knowledge and skills that are applicable to the African context.

Note

1 Kola tenancy law of Eastern Nigeria defines Kola tenancy as the right to use and occupy land by virtue of Kola or other payment made.

References

Abdul-Rasheed, A., Osahon, O., & Alirat Olayinka, A. (2018). Exploring gaps between real estate curriculum and industry needs: A mapping exercise. *Pacific Rim Property Research Journal, 24*(3), 265–283. DOI: 10.1080/14445921.2018.1552472.

Adewunmi, Y. A., & Olaleye, A. (2011). Real estate research directions and priorities for Nigerian institutions. *Journal of Real Estate Practice and Education, 14*(2), 125–140. https://doi.org/10.1080/10835547.2011.12091696.

Black, R., & Rabianski, J. (2003). Defining the real estate body of knowledge: A survey approach. *Journal of Real Estate Practice and Education*, *6*(1), 33–54.

Butler, J. Q., Karl, L. G., & Mimi, W. (1998). Integrating the real estate curriculum. *Journal of Real Estate Practice and Education*, *1*(1), 51–66.

Black, R. T., Carn, N.G., III, J. D., & Rabianski, J. S. (1996). The role of the American real estate society in defining and promulating the study of real property. *Journal of Real Estate Research, 12*(2), 183–193.

Carn, N. G., & Rabianski, J. (1986). Real estate and the AACSB's common body of knowledge. *Real Estate Issues*, *11*(2), 42–48.

Chikafalimani, S. H. P., Thwala, W. D., & Cloete, C. E. (2012, July). Analysis and comparison of masters real estate curricula in South Africa. In *Third international conference on construction in developing countries*, Bangkok, Thailand (pp. 4–6).

Cloete, C. E. (2002). Progress in real estate education in South Africa. *Property Management, 20*(5), 369–382. https://doi.org/10.1108/02637470210450603.

Dasso, J., & Nourse, H. (1992). Body of knowledge for doctoral education in real estate. Paper presented to the American Real Estate Society, San Diego, CA.

Dasso, J., & Woodward, L. (1980). Real estate education: past, present and future – the search for a discipline. *Real Estate Economics*, *8*(4), 404–416.

Dugeri, T. T. (2011). An evaluation of nigerian property market maturity. Unpublished Thesis, Department of Estate Management, University of Lagos, Akoka, Nigeria.

Epley, D. (1996). The current body of knowledge paradigms used in real estate education and issues in need of further research. *Journal of Real Estate Research*, *12*(2), 229–236.

Ezeanya-Esiobu, C. (2019). *Indigenous knowledge and education in Africa, Frontiers in African Business Research*. Springer Open. http://dor.org/10.1007/978-981-13-6635-2_2

Ezema, I., Oluwatayo, A., Adewale, B. and Aderonmu, P. (2015). "Covenant University", available at: http://covenantuniversity.edu.ng/Profiles/Aderonmu-Peter-Adewuyi/Professional-training-of-real-estate-students-A-study-of-academic-curricula-of-three-Universities-in-Nigeria

Ford, D., & Elkes, L. (2008). Team building and communication: keys to success in real estate curricula and the marketplace. *Journal of Real Estate Practice and Education*, *11*(2), 179–185.

Galuppo, L., & Worzala, E. (2004). A study into the important elements of a masters degree in real estate. *Journal of Real Estate Practice and Education*, 7(1), 25–42.

Graeme, N., Elaine, W., Patrick, M., & Karl-Werner, S. (2004). An international perspective on real estate research priorities. *Journal of Real Estate Portfolio Management*, *10*(3), 161–170.

Jayantha, W. M., & Chiang, Y. H. (2012). Key elements of successful graduate real estate education in Hong Kong: Students' perspective. *Journal of Real Estate Practice and Education*, *15*(2), 101–128.

Kampamba, J., Nkwae, B., & Tembo, E. (2015). A comparative analysis of real estate education curricula in botswana. *Mediterranean Journal of Social Science*, 105–116.

Kampamba, J., Tembo, E., & Nkwae, B. (2017). An evaluation of the relevance of real estate curricula in Botswana. *Property Management*, *35*(3), 275–305.

Keogh, G., & D'arcy, E. (1994). Market maturity and property market behaviour: A Euorpean comparison of mature and emerging markets. *Journal of Property Research*, *3*, 215–235.

McCuddy, M. K., Pinar, M., & Gingerich, E. F. (2008). Using student feedback in designing student-focused curricula. *International Journal of Educational Management*, *22*(7) 611–637.

Mirembe, R., & Viruly, F. (2018). Real Estate Knowledge and the Development of a Real Estate Curricula for African Universities. *Journal of African Real Estate Research*, *3*(2), 9–17.

Mooya, M. M. (2007). The impact of property education on property research in South Africa: A review. *RICS Research Paper Series*, 7(6), 9–32.

Musil, T. (2005). Integrating business school curricular resources into real estate practitioner professional development. *Journal of Real Estate Practice and Education*, *8*(1), 133–149.

Nzioki, N., Kariuki, C., & Murigu, J. (2006). *Education and training in valuation, land management and housing in Kenya* (pp. 1–14). www.academia.edu/8048978/Education_And_Training_In_Valuation_Land_Management_And_Housing_In_Kenya.

Olaleye, A., & Adebara, B. O. (2019). Another look at property market maturity framework and its application to Lagos property market, Nigeria. *Journal of Property Investment and Finance, 37*(5), 486–502. https://doi.org/10.1108/JPIF-04-2019-0048.

Poon, J., M., & Hoxley, F. W. (2011). Real estate education: An investigation of multiple stakeholders. *Property Management, 29*(5), 468–487.

Poon, J. A. (2014). Commercial awareness in real estate courses. *Property Management, 32*(1), 48-66. https://doi.org/10.1108/PM-01-2013-0007

Poon, J. B. (2014). Do real estate courses sufficiently develop graduates' employability skills? Perspectives from multiple stakeholders. *Education+ Training, 56*, 562–581.

Rothacher, N. (2013). *Maturity of real estate markets in sub-Saharan Africa.* European Real Estate Society (ERES).

Roulac, S. E. (2002). Requisite knowledge for effective property involvements in the global context. In *Real estate education throughout the world: Past, present and future* (pp. 3–24). Springer.

Schulte, K.-W., & Schulte-Daxbök, G. (2000). Real estate education throughout the world. American Real Estate Society Annual Meeting, Santa Barbara, CA.

Schulte, K.-W. (2001). Interdisciplinary approach of real estate education. PRRES Conference: Adelaide, Australia.

Schulte, K.-W. (2002). Real estate education throughout the world: Past, present and future. In *Research issues in real estate* (Vol. 7). Kluwer Academic Publishers.

Schulte, K.-W., Schulte-Daxbök, G., Holzmann, C., & Wiffler, M. (2005). Internationalisation of real estate education. ERES Conference, Dublin, Ireland.

UN-Habitat (2018). *Valuation of unregistered lands, a policy guide, securing land and property rights for all.* UNON Publishing Services.

Weeks, H. S., & Finch, H. J. (2003). An analysis of real estate curriculum requirements at AACSB international-accredited institutions. *Journal of Real Estate Practice and Education, 6*(2), 257–268.

18

THE REAL ESTATE EDUCATION IN WEST AFRICA

Victor A. Akujuru and Christopher L. Deeyah

1. Introduction

Real estate is a term relating to the physical construction and management of buildings, which is referred to as 'property' in the United Kingdom and other Commonwealth countries. But real estate is now the internationally accepted term, hence its use in this chapter. While real estate studies in the American literature concentrates on the financial management approach and finance studies, the United Kingdom and its former colonies adopt a multi-disciplinary approach with competences covering law, surveying and mapping, building construction and services, business skills and information technology.

The organisation of property markets, real estate management, real estate brokerage and mortgage financing varies greatly between countries. This is reflected in their different types of real estate education. Therefore it is essential for the well-being of any country that those who wish to enter its professions should receive relevant and high quality education. It is arguably obvious of a profession that it should by its very nature seek to demand high standards of its members, and that it should continually and continuously require them to seek to improve their standards, the overall objective being that the profession as a whole should offer an ever improving standard of service to its clientele.

Commercial and industrial activities need the backing of competent professionals in areas such as law, banking and property professions, even as competent professional people also have much to contribute to society in respect of social and ideological issues. It is therefore necessary that courses designed for the various professions should produce the right sort of professionals for conditions in the contemporary world, which are very different from those of the only the relative recent past (Millington, 2003). Small and Karantonis (2001) in their study of the challenges of balancing the provisions of education and training within vocational courses in land economy recommended that property courses should incorporate both the classroom as well as practical training, for the university students to be technically sound and well-equipped to respond to challenges of change. **Real estate education has a big role to play in the production of professionals to meet the standard of what the society demands**. In order to appreciate the structures of real estate education in the West

DOI: 10.1201/9780429279256-18

African sub-region, it is necessary to have some understanding of its institutional arrangements particularly in Nigeria and Ghana.

Real estate education in Nigeria and Ghana has been in existence for a long time. Since inception, no curriculum evaluation was conducted to compare subjects that are offered in real estate curricula in both countries at a local level as well as with similar curricula internationally. In Nigeria, the Rivers State University RSU, Port Harcourt runs the Bachelor of Technology B.Tech degree in estate management, while in Ghana the University of Science and Technology UST Kumasi runs a four year B.Sc degree in land economy. Both programmes are offered at university level in Nigeria and Ghana and are accredited by their respective regulatory body. The real estate programmes are known as estate management in Nigeria and land economy in Ghana.

Like in other countries, real estate education in West Africa is very different owing to varying culture, custom and tradition as well as national land laws. This study is about analysing the curriculum design in Anglophone West Africa with a view to identify the differences and similarities in the real estate education curricula and to suggest areas needing improvements to meet current trends in globalisation. Shelton and Howard (2003) and Kampamba et al. (2015) mentioned that the purpose of analysing real estate curricula is (i) to provide support for educators who are developing their curricula; (ii) to extend the real estate literature; and (iii) to provide factors to be included when developing or reviewing real estate programmes. The chapter's objective therefore is to carry out a comparative analysis of the programmes offered at degree level at the Rivers State University RSU Port Harcourt, Nigeria and the University of Science and Technology UST Kumasi in terms of curriculum structure and to establish the level of similarity between these curricula in real estate education; the difference within the curricula in real estate education that are being offered in the two countries; and the level of standardisation as far as real estate education curricula in West Africa is concerned.

Following the introduction, there is a review of literature on real estate education body of knowledge, and real estate training and education. This is followed by the overall programme philosophy, career opportunities in Nigeria and Ghana. Thereafter, education and training in real estate of both countries are evaluated as a foundation for comparative analysis of curricula content. The data on the findings of the comparative analyses of courses offered in both countries were presented and recommendations of other reviewed works are then collated from literatures on standardisation in real estate education curriculum.

2. Literature Review

Numerous scholars have observed that there is no universally accepted real estate body of knowledge (Boyd et al., 2013; Chikafalimani, 2010; Dasso & Woodward, 1980; Jay, 2011; Schulte et al., 2005; Yu, 2001; Shelton & Howard, 2003). In trying to explain why real estate approaches vary in different geographic locations, Mooya (2007) said real estate developed differently in various nations. In USA, it evolved from the finance discipline while in the UK, it evolved from the land economics discipline. Boyd et al. (2013) and Yu (2001) reasoned that the lack of consensus in the real estate body of knowledge is caused mainly by the complex nature of property and by its texture which is ever changing. Mooya (2007) further opined that the differences in real estate paradigms produce graduates with different ways of understanding and problem solving techniques and methods. Dasso and Woodward (1980) shared the similar sentiments arguing that lack of consensus in the real estate body of knowledge affected the development and management of real estate programmes.

Studies have also examined real estate training and education. A study by Poon et al. (2011) evaluated real estate stakeholders' perspectives of gaps between the employers' expectations of real estate graduates and what real estate graduates considered to have attained during their studies in line with the content of Royal Institution of Chartered Surveyors (RICS) accredited real estate courses. The authors found that practical experience is missing from university courses but made no conscious efforts to show how the identified gaps can be filled.

Small and Karantonis (2001) considered the challenges of balancing the provision of education and training within vocational courses in land economy. They recommended that property courses should incorporate both the classroom education as well as practical training, to make university students technically sound and well equipped to respond to challenges of change or to the interpretation of the relationship between his or her occupation and the environment within which he works. The authors however failed to show, in practical terms, how to bridge perceived gaps in real estate education.Crews (2004) was motivated by the challenge of the scarcity of professional real estate sales persons resulting from lack of interest by salesmen in advanced real estate education to examine the effectiveness of real estate programmes in creating a robust academic and adult learning environment capable of meeting the needs of the students who require up-skilling but have limited time for study. The author observed a gap in skill and recommended innovative and flexible modes of learning suitable for a nationwide, busy work force but failed to show how the identified gap could be filled.

There are lots of skills that have shown to be missing from the real estate curriculum design. Black et al. (1996) noted that some basic skills and knowledge, such as negotiation skills, information gathering and processing, management of people and processes, communication, problem solving and knowledge about market evaluation and business environment, required of an estate surveyor are, in most cases, not learnt in school.

Oloyede et al. (2017) evaluated the gaps in the programme of study of Departments of Estate Management in Nigerian universities to determine the extent to which the needs of the property industry are being met by the current curriculum using secondary data from the review of current literature. The study found that there were significant differences in the views of researchers across the globe concerning practical experience, graduates' levels of commercial awareness, information technology tools, negotiation skills, information gathering and processing, management of people and processes, communication, problem solving and knowledge about market evaluation. The study recommends real estate management programmes to include adventure education courses geared more towards the promotion of soft skills in students.

3. Programme Philosophy

Estate management programmes are designed to produce competent appraisers/ estate surveyors and valuers with sufficient technical knowledge and skills to optimise the use of land resources and hereby facilitate the economic development of the nation. Estate management is concerned with the supervision and control of an interest in landed property (Thorncroft, 1965). It may also be defined as the study of the use, development and management of land, other natural resources and the built environment. It is called by different names in different countries.

The fundamental task of the appraiser/ estate surveyor and valuer is to obtain the maximum economic returns from available resources, having regard to the social advantages and long-term economic developments. The course is designed to prepare students for responsible career positions in the land management profession upon graduation. The varying demands and complex legal, socio-economic and political factors relating to land calls graduates with considerable understanding and in-depth knowledge of the theoretical base relating to the administration of the land, together with a sound and broadly based knowledge of the techniques and practices of property valuation, building construction, land surveying, town planning, land law, land economics property management and property development.

There is a wide range of opportunities in occupations dealing with land management, development and planning, property valuation, estate agency, property maintenance, etc. Others find interesting careers in banking and other financial institutions, in consultancy, in management, agriculture and in urban and regional analysis.

4. Career Opportunities

Graduates of real estate have a number of job opportunities open to them. Any person or any organisation that occupies, owns, seeks to own or finances property is likely to require the service of an estate surveyor and valuer. This can broadly be categorised into two sectors as:

1. Private sector
   - i. private individuals
   - ii. insurance companies
   - iii. financial institutions like pension funds
   - iv. financial institutions like commercial banks, merchant banks, development banks and government banks
   - v. industrial and commercial companies
   - vi. property companies
   - vii. property developers and construction companies
   - viii. overseas buyers
2. Public sector
   - i. government establishments such as property development authorities, ministry responsible for physical development, land administration and taxation
   - ii. local authorities
   - iii. public investment corporations
   - iv. federal government departments and agencies

The services of an estate surveyor and valuer can thus be used by anyone with a legal interest in landed property or considering acquiring such an interest. Potential clients can range from major global organisations down to individual citizens. Large numbers of valuers work in private practice for large or small property consultancy or surveying firms, in partnerships or as sole practitioners, and act for clients. Alternatively, many large organisations, including some sections of central government, have their own 'in-house' valuers who act for them on their property matters. There is thus a very wide scope of potential employers. Geographically there are opportunities on a worldwide scale with RICS qualified professionals being recognised internationally.

5. Research Method

The research method adopted in this study is documents review of extant literature and on real estate education. A total of 11 publications on the topic were found through online search. This was found to be a reasonable method of assembling published work on the topic. The papers reviewed are empirical studies undertaken between 1996 and 2018, a period spanning 22 years. The documents which were also reviewed are basically department handbooks which provided reliable information on the curricula of the Rivers State University, Port Harcourt, Nigeria and the University of Science and Technology, Kumasi, Ghana. The document review was particularly useful as it enables trends to be studied easily and data analysis is meaningful. Such documents prove valuable not only because of what can be learned directly from them but also stimulus for paths of inquiry that can be pursued only through direct observation.

6. Nigerian Education and Training in Real Estate

Independent in 1960 the Federal Republic of Nigeria was a former British colony. It is located on the Atlantic coast of West Africa. The federal capital is Abuja and the commercial capital is Lagos. Nigeria is the most populous country in Africa, with an estimated 299 million people. Exports are mainly crude oil and cocoa (Ghyoot, 2002).

Historically, the profession of estate management in Nigeria came in the wake of colonialism with F.G Gleave, an expatriate being the first known chartered surveyor to set up a general practice of estate surveying and valuation firm in the country. Estate management as an academic discipline began in 1957, when the then Nigeria College of Arts, Science and Technology (now university of Nigeria Nsukka) offered the Royal Institution of Chartered Surveyors syllabus to the immediate examination after which the students had to be sent to England. That course metamorphosed into the Estate Management department of the university of Nigeria in 1962 (Umeh, 1986). Today, Nigeria has more than 31 universities and 48 polytechnics in Nigeria offering courses in estate management due to the increasing need and interest in the real estate profession.

The real estate practice is regulated by Decree No. 24 of 1975 now an Act under CAP 111, Laws of the Federal republic of Nigeria, which defines the profession as:

Those engaging in the art, science and practice of:

i. Determining the value of all descriptions of property and the various interests therein
ii. Managing and developing estates and other business concerned with the management of landed property
iii. Securing the optimal use of land and its associated resources to meet social and economic needs
iv. Determining the structure and condition of buildings and their services and advising on their maintenance, alteration and improvement

v. Determining the economic use of those resources by means of financial appraisal for the building industry
vi. Selling (whether by auction or otherwise), buying or letting as an agent, real or personal property or any interest therein (Source: *NIESV Code of Professional ethics and practice*)

This law established the Estate Surveyors and Valuers Registration Board of Nigeria (ESVARBON) as the regulatory organ of government, to licence and regulate the practice of estate surveying and valuation in Nigeria. Professionals in this field of endeavour all belong to the Nigerian Institution of Estate Surveyors and Valuers (NIESV) which is the professional association of practitioners of estate surveying and valuation. Those to be registered by the Board must first qualify to be and be registered by the professional association, which then presents such applicants to the Board for registration. There are three routes or paths to becoming an estate surveyor and valuer. The first involves undertaking a five-year programme in an approved university leading to the award of a degree in estate management while the second involves taking diploma courses in the polytechnics. The diploma course is broken into the National Diploma (ND) and a Higher National Diploma (HND) consisting of two years respectively. There is a one-year unsupervised industrial training upon completion of the National Diploma programme. Additionally, there is also a four-month supervised industrial training after the first year of the National Diploma programme. The third path involves a candidate undertaking practical training in an accredited estate surveying and valuation firm while at the same time sitting for the professional qualifying examinations of the Nigerian Institution of Estate Surveyors and Valuers (NIESV).

6.1 University Under-Graduate Programme Description and Core Curriculum

At the undergraduate level, real estate is mostly taught in the Department of Estate Management, in the Faculty of Environmental Sciences, Rivers State University. The Department of Estate Management, of the Rivers State University, Port Harcourt, Nigeria has a five-year B.Sc (Estate Management) programme which runs for 10 semesters, with the students being exposed to Supervised Industrial Work Experience (S.I.W.E.) during the second semester of the fourth year.

6.2 Aim and Objectives of the Programme

The aim of this programme as shown in table 18.1 is to provide fundamental knowledge in, and deepen the students' grasp of land use and land resource management and property valuation; it develops in them the spirit of inquiry and generally prepares them to face challenges and solve problems in the field of estate management and valuation.

The objectives of the programme are to:

i. Instil in students an appreciation of the complex nature of the relationships between man and his environment.
ii. Involve the students in an intellectually stimulating and satisfying experience of learning and studying.

TABLE 18.1 Syllabus for the B.Tech (Estate Management) (Department of Estate Management, Rivers State University RSU, Nkpolu- Oroworukwo, Port Harcourt, Nigeria)

Year 1: First Semester

ERD 111 Introduction to General Agriculture
ARC 111 Architectural Graphics 1
ECO 111 Principles of Economics 1
EMS 111 Introduction to Estate Management 1
GST 141 Use of English 1
GST 130 Introduction to African History
MTH 111 Algebra and Trigonometry
SVG 111 Basic Surveying 1
URP 111 Introduction to the Basic Elements of Urban and Regional Planning

Year 1: Second Semester

GST 112 Man and Society
MTH116 Calculus
GST 142 Use of English II
ARC 112 Architectural Graphics II
SVG 112 Basic Surveying II
ECO 102 Principles of Economics II
EMS 112 Introduction to Estate Management II
URP 111 Urban Development Planning

Year 2: First Semester

EMS 211 Introduction to Valuation
LAW 24 Law of Contract and Tort 1
URP 261 Nature of Environmental Sciences
GST 150 Philosophy and Logic
QUS 111 Principles of Quantity Surveying I
ARC 261 Building Construction and Materials I
MTH 161 Introductory Statistics
LAW 113 General Principles of Nigerian Law
EDC 211 Introduction to Entrepreneurship Studies

Year 2: Second Semester

EMS 222 Principles and Methods of Valuation I
LAW 242 Law of Contract & Tort II
GST 250 History and Philosophy of Science
QUS 112 Principles of Measurement & Description
ARC 262 Building Construction and Material II
EMS 242 Land Use Economics I
URP 252 Planning Laws & Procedure
GST 222 Peace Studies & Conflict Resolution

Year 3: First Semester

EMS 323 Principles and Methods of Valuation II
LAW 331 Land Law I
EMS 343 Land Use Economics II
ARC363 Building Construction and Materials III
EMS 313 Arbitration and Award I
ACC 314 Estate Accounts I
EMS 315 Environmental Management

Year 3: Second Semester

EMS 324 Advanced Valuation I
LAW 332 Land Law II
CMS 102 Introduction to Computer Science
MTH 262 Statistical Theory & Applications
ACC 315 Estate Accounts I
QUS 342 Construction and Building Economics I
EDC 310 Entrepreneurial Practice
EMS 344 Urban Land Use Policy

Year 4: First Semester

EMS 425 Advanced Valuation II
EMS 445 Information Technology in Rea Estate Practice
EMS 471 Property Development
EMS451 National & Local Taxation
EM 435 Facility Management
EMS 433 Management Practice
EMS 461 Landlord and Tenant Relationship
CMS 211 Computer Programming and Applications

Year 4: Second Semester

EMS 491 Student's Industrial Work Experience Scheme (SIWES)

Year 5: First Semester

EMS 527 Applied Valuation I
EMS 571 Feasibility and Viability Studies
EMS 537 Building Maintenance & Services I
EMS 581 Research Techniques/Seminar
EM 529 Environmental Valuation
EMS 539 Applied Property Management
EMS 553 Mineral Valuation

Year 5: Second Semester

EMS 528 Applied Valuation II
EMS 538 Building Maintenance & Services II
EMS 516 Professional Practice and Code of Conduct
EMS 582 Project
EM 554 Plant Machinery Valuation
EMS 566 Project Management
EMS 564 Housing Finance

Source: Department of Estate Management Handbook, Rivers State University (RSU), 2018.

iii. Provide a broad understanding on buildings, their construction and use.
iv. Develop in students the ability to apply analytical skills to finding solutions to theoretical and practical land resources problems.
v. Develop in the students survival skills in an ever-changing economic, technological and political world.
vi. Lay the foundation on which students can proceed to further studies in specialized aspects of estate surveying practice, of multi-disciplinary areas involving estate surveying and valuation.
vii. Create an appreciation of the importance of estate surveying and valuation in an industrial, environmental, economic and social context.
viii. Develop the students in the use of information technology in the effective management of land and the environment (*Department of Estate Management Handbook Rivers State University, Nigeria*, 2018).

6.3 Entry Requirements

Admissions into the Bachelor of Science (B.Sc)/Bachelor of Technology (B.Tech) degree programme in estate management are normally through the Joint Admissions and Matriculation Board (JAMB).

I. Admission through UME

Candidates shall be admitted into year 1 of the degree programme where he/she obtains in not more than two sittings, credit or higher level passes at the Senior School Certificate SSC, NECO examination in a minimum of five subjects including mathematics, English language and economics. The other two subjects shall include geography, biology, agricultural science and other relevant science subjects.

II. Admission to Year II

a. HSC/GCE 'A' level passes in at least two relevant subjects (one of which must be economics) plus school certificate/GCE 'O' level credit passes (including English language and mathematics) in three other subjects from those indicated earlier, at not more than two sittings.
b. Candidates who passed with at least Upper Credit in the National Diploma (ND) from recognised institutions may be admitted into the second year degree programme provided such candidates also satisfy the requirements for admission through UME.

III. Admission to Year III

Candidates who passed with at least Upper Credit in the Higher National Diploma (HND) are eligible for admission into the third year of the programme, provided that such candidates also satisfy the requirements for admission through UME.

6.4 Polytechnic Programme Description and Core Curriculum

The polytechnic education was established to provide training and research in engineering, applied sciences, environmental technology, business management and such other fields of learning as may be determined from time to time with regard to the work force needs of the state in particular and of the federation in general especially the development of middle-level manpower. At the polytechnic level, real estate is taught in the school of environmental technology. The estate management programme is offered at the National Diploma (ND) and the Higher National Diploma (HND) levels in the School of Environmental Technology. The Department of Estate Management, Kenule Beesor Saro-Wiwa Polytechnic, Bori, and Rivers State, Nigeria is established to acquaint the students with sufficient and relevant knowledge and skills that would lead them to the practice of estate management and valuation. The core courses and the service courses offered by the department are contained in the approved National Board for Technical Education Curriculum and Course Specification in Estate Management and the polytechnic local content courses as shown in table 18.2.

6.5 Graduate Programmes in Real Estate

In recent times, it has become necessary to provide specialist training in real estate studies, different from the generalist 'jack-of-all-trade' type of training offered at the undergraduate levels. There are now graduate programmes at the Post Graduate Diploma (PGD), Master of Science (MSc) and the Doctor of Philosophy (PhD) levels with specialisations in various aspects of real estate, land economy and land governance. Candidates enrolled in the graduate programmes have various areas of specialisations like:

Investment Analysis and Valuation;
Investment Appraisals and Development;
Environmental Valuations and Management;
Facility and Property Management;
Land Governance and Sustainable Development; and
Housing Studies.

While the aim is to offer opportunities for in-depth research in different aspects of real estate and provide qualified manpower to train real estate professionals locally, the programmes aim to:

Facilitate the development of critical and logical skills for high quality research in real estate;
Facilitate independent and original research within an interdisciplinary environment necessary for the development of scholarly skills and specialisation;
Undertake real estate research that has significant contribution to the advancement and understanding of real estate markets efficiency and knowledge;
Promote high quality scholarship in areas of critical importance in the field of real estate;
Produce experts in the field for high level academic positions in Institutions of Higher Learning; and
Produce intellectual leaders of thought capable of breaking new frontiers of knowledge in all aspects of real estate.

Entry into the graduate programmes generally requires excellence at the undergraduate levels with the minimum CGPA of 3.00 at the PGD and 3.00 at the MSc levels to enroll at the MSc and PhD programmes respectively, in addition to the possession of a cognate degree/qualification with and Upper Credit or Second Class Lower levels of pass at the HND and BSc programmes respectively.

TABLE 18.2 Syllabus for the ND/HND (Estate Management) (Department of Estate Management and Valuation, Kenule Beesor Saro-Wiwa Polytechnic, Bori, Rivers State, Nigeria)

National Diploma Curriculum	
ND 1: First Semester	**ND 2: First Semester**
GNS 111 Use of English I	GNS 201 Use of English 2
BFN112 Principles of Economics 1	STA 111 Introduction to Statistics
GNS 118 Citizenship Education	URP 117 Urban and Regional Planning Principles and Techniques
MTH 111 Mathematics	
EST 111 Introduction to Estate Management	EED 126 Introduction to Entrepreneurship
EST 112 Introduction to Accounting 2	BLD 101 Building Construction I
EST 113 Nigerian Legal systems	EST 211 Principles of Valuation I
MEC111 Technical Drawing 1	EST 212 Land Law I
SUG 111 Principles in Surveying I	EST 213 Introduction to Property Development
PHE 111 Physical and Health Education	EST 214 Rural Land Use Economics
	EST 215 Property Taxation
	EST 216 Research Methods
ND 1: Second Semester	**ND 2: Second Semester**
BFN122 Principles of Economics 2	GNS 221 Communication in English II
GNS 121 Communication in English II	EED216 Practice of Entrepreneurship
GNS 121 Introduction to Sociology 2	BLD 102 Building Construction II
MTH 121 Mathematics 2	EST 221 Principles of Valuation II
EST 122 Principles of Accounting 2	EST 222 Land Law II
EST 123 Law of Contract and Torts	EST 223 Introduction to Property Management
EST 124 Introduction to Valuation	EST 224 Urban Land Use Economics 2
COM 111 Introduction to Computer	EST 225 Principles of Property Rating
SUG 102 Principles in Surveying II	EST 226 Building Services and Maintenance
EST 122 Estate Graphics	EST 227 Final Project
Higher National Diploma Curriculum	
HND 1: First Semester	**HND 2: First Semester**
GNS 301 Use of English	GNS 411 Introduction to Psychology
BLD 203 Building Construction III	EST 418 Computer Application to Real Estate
EST 315 Building Economics	EST 411 Advance Valuation I
EST 311 Valuation 1	EST 413 Property Management I
EST 313 Property Development 1	EST414 Property Rating & Taxation
EST 316 Estate Office Administration	Est415 Real Estate Appraisal I
EST312 Property Law 1	EST 416 Professional Practice
EST 314 Land Economics 1	EST 417 Professional Report Writing
EST 320 Facilities Management	EST 412 Arbitration & Award
EST 317 Environment & Estate Services	EST 419 Mineral Valuation
HND 1: Second Semester	**HND 2: Second Semester**
GNS 302 Communication in English	EST 421 Advance Valuation II
BLD 203 Building Construction IV	EST 423 Property Management II
GNS323 Psychology	EST424 Property Rating & Taxation II
EST 321 Valuation II	EST 425 Real Estate Appraisal 2
EST 323 Property Development II	EST 422 Compulsory Acquisition & Compensation Law
EST322 Property Law II	EST 426 Project Dissertation
EST 324 Land Economics 2	EST 427 Plant & Machinery Valuation
EST 327 Estate Services	EST 429 Environmental Valuation
EST 325 Estate Accounting	EST 428 Environmental Impact Assessment
EST 326 Research Methodology	
URP 128 Urban Principles and Techniques	

Source: NBTE Curriculum (2002) and Department of Estate Management Local Content Courses.

7. Ghanaians Education and Training in Real Estate

Another West Africa country independent from Britain since 1957 that is training real estate practitioners is the Republic of Ghana located along the Gulf of Guinea and Atlantic Ocean, in the sub-region of West Africa. The current population is 30,554,589 (2016) with English as the official language. The capital is Accra. Major industries are mining, light manufacturing, aluminum smelting and food processing. It is the largest producer of cocoa in the world and earns most of its earnings from it (Ghyoot, 2002).

To become a real estate surveyor in Ghana, academic training in valuation and estate management is a perquisite. Graduates in valuation or estate management must possess property-related degrees or Higher National Diploma (HND) in land economics or estate management respectively. The Department of Land Economy of Kwame Nkuruma University of Science and Technology awards the degree, while the Department of Estate Management of Kumasi Polytechnic awards the National Diploma in estate management.

7.1 Under-Graduate Programme Description and Core Curriculum

The University of Science and Technology UST Kumasi has been presenting a four-year B.Sc degree in land economy since. The syllabus for the programmes are designed to equip students with the necessary theoretical knowledge with a strong complement of practical training in the real estate industry, to enable students to meet the demands of the local industry and cope with modern trends in the real estate industry.

7.2 Aim and Objectives of the Programme

The programme is aimed at equipping students with the requisite knowledge and skills in harnessing the potentials of land resources in delivering optimum returns to individuals, groups and society at large.

The key objectives of the training include:

i. The assessment of value of property for various purposes
ii. The administration and management of urban land and buildings
iii. Assess the feasibility and viability of appraisal of proposed landed projects
iv. The role of 'play maker' in the property market in the mediation between suppliers and consumers
v. The planning, development and redevelopment of property schemes as client's representative
vi. Develop innovative instruments for ownership of real estate products
vii. Identify and resolve land ownership and use conflicts
viii. Manage landed property in environmentally friendly ways
ix. The monitoring of all land based activities in so far as they affect the best use of land and environmental impact assessments
x. Effectively liaise with other professional members of the development team in the design and construction of real estate products

The syllabus covers the theory and methodology of valuation, planning and research methods. Table 18.3 provides more details.

TABLE 18.3 Syllabus for the Four-Year B.Sc in Land Economy (Department of Land Economy and Estate Management, University of Science and Technology, Kumasi, Ghana)

Year 1: First Semester	**Year 3: First Semester**
LE 151 General Principles of Law	LE 351 Law Of Property I
LE 153 Building Construction I	LE 353 Advanced Valuation I
LE 155 Land Economics I	LE 355 Land Use Planning & Administration I
CSM183 Introduction to Computers I	LE 357 Industrial Structures and Quantities I V
LE 159 Financial Mathematics I	LE 359 Principles of Accounting I
ENGL 157 Communication Skills	LE 361 Research Methods
GE 183 Principles of Surveying	LE 363 Estate Management 1
Year 1: Second Semester	**Year 3: Second Semester**
LE 152 Law of Contract	LE 352 Law of Property II
LE 154 Building Construction II	LE 354 Advanced Valuation II
LE 156 Land Economics II	LE 356 Land Use Planning & Administration II
CSM 184 Introduction to Computers II	LE 358 Industrial Structures and Quantities II
LE 160 Financial Mathematics II	LE 360 Principles of Accounting II
ENGL 158 Communication Skills	LE 362 Research Methods II
LE 162 Principles of Forestry	LE 364 Estate Management II
GE 164 Introduction to Geo-Information Systems	
Year 2: First Semester	**Year 4: First Semester**
LE 251 Law of Tort	LE 451 Law Relating to Estate Management I
LE 253 Principles of Valuation I	LE 453 Estate Agency
LE 255 Introductory Statistics I	LE 455 Ethics and Professional Practice I (2–2–3)
LE 257 Building Construction and Services I	LE 457 Real Estate Finance I
LE 259 Elements of Central and Local Government I	LE 459 Rating and Taxation I
LE 261 Fundamentals of Planning	
ENGL Literature in English I	
Year 2: Second Semester	**Year 4: Second Semester**
LE 252 Principles of Customary Land Law	LE 452 Law Relating to Estate Management II
LE 254 Principles of Valuation II	LE 456 Ethics and Professional Practice II (2–2–3)
LE 256 Introductory Statistics II	LE 458 Real Estate Finance II
LE 258 Building Construction and Services II	LE 460 Rating and Taxation II
LE 260 Elements of Central and Local Government II	LE 492 Project Work
LE 262 Fundamentals of Planning II	
ENGL Literature in English II	

Source: Information for Programme Re-Accreditation Application of BSc. Land Economy May 2016. Kwame Nkrumah University of Science and Technology, Kumasi – Ghana.

7.3 Polytechnic Programme Description and Core Curriculum

The polytechnics in Ghana were first established as technical institutes that offered craft courses. In 1960, following the industrial development policy and rapid technological progress in a broad range of areas, technical education became a necessity for the country. Since the technical institutions (polytechnics) were offering second-cycle craft courses while the universities were offering higher tertiary courses, there was a gap in the manpower supply needs of the country (Nsiah-Gyabaah, 2005). In recognition of this, a number of the technical institutes were established to train lower and middle-level skilled manpower to fill the gap. As part of the tertiary reform programme the polytechnics were directed to give full implementation to the reform. This involved the mounting of Higher National Diploma (HND) programmes in various fields. The White Paper of the Ministry of Education expressly confirmed that the Polytechnics have a distinct and important role to play in middle-level manpower development, and that programmes and courses were to be offered at the middle-level of technical training leading to the award of higher national diplomas but not departing from syllabi dedicated to practical training. The provision of such programmes will complete the cycle of technical education and provide a capacity for higher level technician training and practical research (Ministry of Education, 1993). Simultaneously, the universities were to wind up their diploma programmes. Pursuant to this, the Kumasi Polytechnic started preparing the grounds for mounting the HND programmes including the HND in estate management. Unlike some other disciplines in the polytechnic, estate management was new as such; a new department had to be established to run the programme. The programme started at the beginning of February 1996. These technical institutes were established in Accra, Kumasi and Takoradi (Boakye-Agyeman, 2006).

In the case of the surveying profession, technicians are indispensable in the collection and analysis of field data required for decision making. Since polytechnic education emphasises technical training, it is imperative that they offer technical programmes in surveying. All the polytechnics in the country run programmes in building technology. Technical training in estate management is offered only at the Kumasi Polytechnic, thus giving it a national character. Currently, the department runs a three-year (six semesters) HND programme in estate management. Students take courses in Department of Estate Management, Kumasi Polytechnic.

8. The Profession of Real Estate Practitioners

The real estate profession is currently regulated in Anglophone West Africa. In Nigeria, professionals belong to the Nigerian Institution of Estate Surveyors and Valuers (NIESV) but are regulated by the Estate Surveyors and Valuers Registration Board of Nigeria (ESVARBON). The Institution is a professional body and regulates its members; organises national conferences, seminars and workshops; and sets professional examinations for prospective estate surveyors and valuers. It also administers the continuing professional development (CPD) programmes in conjunction with the Registration Board.

In Ghana, real estate practitioners are known as valuers and its founders are those who received their training in the United Kingdom and became members of the Royal Institution of Chartered Surveyors (RICS). These founding fathers together with their counterparts from the Quantity and Land surveying profession constituted the then Ghana Branch of the Royal Institution of Chartered Surveyors (RICS) professionals. However in 1969, the Ghana Branch of the RICS at an Annual General Meeting established the Ghana Institution of Surveyors (GhIS). The GhIS since then has been responsible for the regulation of general surveying practice including valuation for its members. The University of Science and Technology now Kwame Nkuruma University of Science and Technology has been the main initial training ground for members of GhIS (Sawyer, 1972). The GhIS is an umbrella organisation for all allied professionals in the built environment, and it is structured along the lines of the RICS of Great Britain.

TABLE 18.4 Syllabus for the Three-Year HND Estate Management (Department of Estate Management, Kumasi Polytechnic, Ghana)

Year 1: First Semester	**Year 2: Second Semester**
ESM 101 Law I – Principle of Law & Equity	ESM 202 Law Relating to Estate Mgt II
ESM 103 Elements of Central & Local Government	ESM 204 Principles and Practice of Valuation II
ESM 105 Principles of land Surveying I	ESM 206 Land Economics II
ESM 107 Principles Accounting I	ESM 208 Building Construction IV
ESM 109 Mathematics	ESM 210 Land Use Planning & Administration II
ESM 111 Building Construction I	ESM 212 Property Management II
ESM 105 Principles of Economics	ESM 216 Rating and Taxation I
CLT 101 Communication Skills I	ETP 202 Entrepreneurship II
CLT 111 Computer Literacy I	
Year 1: Second Semester	**Year 3: First Semester**
ESM 102 Law II – Contract & Tort	ESM 301 Practical Training in Industry
ESM 104 Elements of Central & Local Government	ESM 303 Technical Report
ESM 106 Principles of Land Surveying II	ESM 305 Project Work
ESM 108 Principles of Accounting II	
ESM 110 Introduction to Valuation	
ESM 112 Building Construction II	
Econ 106 Principles of Economics II	
COS 102 Communication Skills II	
CLT 111 Computer Literacy II	
AS 100 African studies	
Year 2: First Semester	**Year 3: Second Semester**
ESM 201 Law Relating to Estate Mgt I	ESM 302 Law Relating to Estate Management V
ESM 203 Principles and Practice of Valuation I	ESM 304 Principles & Practice of Evaluation III
ESM 205 Land Economics I	ESM 306 Rating & Taxation II
ESM 207 Building Construction III	ESM 308 Building Construction V
ESM 209 Land Use Planning & Administration I	ESM 310 Elements of Building Quantities
ESM 211 Property Management I	ESM 312 Property Management III
ESM 213 Research Methods	
ETP 201 Entrepreneurship I	

Source: Boakye-Agyeman N. A. (2006). Polytechnic Education in Ghana: The Case of the HND Estate Management Programme.

9. Data Analysis – Comparison of Real Estate Education Curricula in West Africa

This section presents results from documents reviewed on the comparison of real estate education curricula on courses offered in real estate degree programmes and those offered at the diploma programmes of four selected schools in West Africa. The schools are the Rivers State University, Nkpolu-Oroworukwo, Port Harcourt, Nigeria; University of Science and Technology, Kumasi – Ghana; Kenule Beesor Saro-Wiwa Polytechnic, Bori, Rivers State, Nigeria; and the Kumasi Polytechnic, Ghana. The differences and similarities of degree and diploma in real estate education curricula in terms of topics covered are tabulated and discussed under this section. The results are shown in Tables 18.5 and 18.6; and Figures 18.1 and 18.2:

TABLE 18.5 Comparison of Courses Offered in Real Estate Degree Programmes

S/No.	*Courses offered*	*RSU Port Harcourt, Nigeria*	*UST Kumasi, Ghana*
1	Advanced Valuation	√	√
2	Algebra and Trigonometry	√	X
3	Applied Property Management	√	X
4	Applied Valuation	√	√
5	Arbitration and Award	√	X
6	Architectural Graphics	√	X
7	Basic Surveying	√	√
8	Building Construction and Materials	√	√
9	Building Maintenance & Services	√	√
10	Calculus	√	X
11	Communication in English	√	√
12	Computer Programming and Applications	√	X
13	Construction and Building Economics	√	X
14	Elements of Central and Local Government	X	√
15	Entrepreneurial Practice	√	X
16	Environmental Management	√	X
17	Environmental Valuation	√	X
18	Principles of Accounting	√	√
19	Estate Agency	X	√
20	Introduction to Estate Management	√	√
21	Ethics and Professional Practice	√	√
22	Facility Management	√	X
23	Feasibility and Viability Studies	√	X
24	Financial Mathematics	X·	√
25	General Principles of Law	√	√
26	History and Philosophy of Science	√	X
27	Housing Finance/Real Estate Finance	√	√
28	Industrial Structures and Quantities	X	√
29	Information Technology in Real Estate Practice	√	X
30	Introduction to African History	√	X
31	Introduction to Computer Science	√	√
32	Introduction to Entrepreneurship Studies	√	X
33	Introduction to General Agriculture	√	√
34	Basic Elements of Planning	√	√
35	Introduction to Geo-Information Systems	√	X
36	Introduction to Valuation	√	X
37	Introductory Statistics	√	√
38	Land Economics	√	√
39	Land Law	√	√
40	Land Use Planning & Administration	X	√
41	Urban Development Planning	√	X
42	Urban Land Use Policy	√	X
43	Landlord and Tenant Relationship	√	X
44	Law of Contract and Tort	√	√
45	Law Relating to Estate Management	X	√
46	Literature in English	X	√
47	Man and Society	√	X
48	Management Practice	√	X

S/No.	*Courses offered*	*RSU Port Harcourt, Nigeria*	*UST Kumasi, Ghana*
49	**Mineral Valuation**	√	X
50	**Rating and Taxation**	√	√
51	**Nature of Environmental Sciences**	√	X
52	**Peace Studies & Conflict Resolution**	√	X
53	**Philosophy and Logic**	√	X
54	**Planning Laws & Procedure**	√	X
55	**Plant Machinery Valuation**	√	X
56	**Student's Industrial Work Experience Scheme (SIWES)**	√	X
57	**Principles of Economics 1**	√	X
58	**Principles of Valuation**	√	√
59	**Project Work**	√	√
60	**Project Management**	√	X
61	**Property Development**	√	X
62	**Research Methods**	√	√
63	**Statistical Theory & Applications**	√	X

Legend: √ = Course offered; X = Course not offered

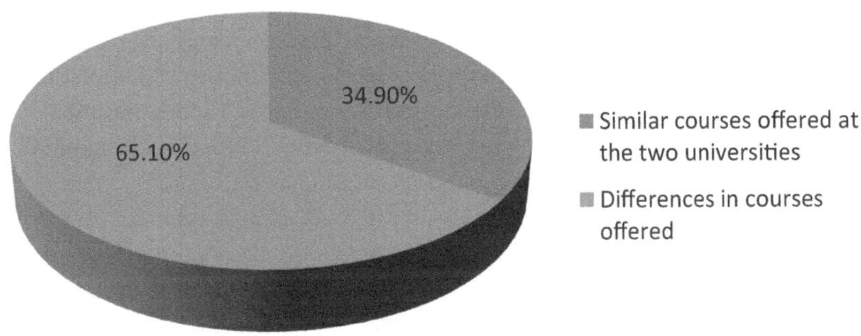

FIGURE 18.1 Comparison of University Course Curriculum in Nigeria and Ghana

9.1 Comparison of Real Estate Education University Curricula in Nigeria and Ghana

This section provides a comparative analysis of real estate education curriculum in Nigeria and Ghana. Table 18.5 presents the degree curricula for real estate offered at the two universities. A total of 63 courses was identified as being offered in two semesters.

Table 18.5 shows a difference of 41 (65.10%) and a similarity of 22 (34.90%) in the course curriculum offered by the two universities in West Africa. This is illustrated in figure 18.1. The differences in the two programmes imply non-standardisation of courses at the West African countries.

9.2 Comparison of Polytechnic Curricula in Real Estate Education in Nigeria and Ghana

This section provides a comparative analysis of polytechnic curricula in real estate education in Nigeria and Ghana. Table 18.6 presents the diploma curricula for real estate offered at the two polytechnics. A total of 56 courses were identified as being offered in two semesters.

Table 18.5 shows a difference of 35 (62.50%) and a similarity of 21 (37.50%) in the course curriculum offered by the two universities in West Africa. This is illustrated in figure 18.2. The differences in the two programmes further imply that there is non-standardisation of courses in the West African countries.

9.3 The Need for Standardisation of Real Estate Curriculum

Practitioners in both Ghana and Nigeria are optionally affiliated to RICS, and both countries run academic curricula at both the undergraduate and postgraduate levels similar to that of RICS syllabus, therefore, differences noticed in the curriculum of both countries is least expected. It is a truism that this non-standardisation has not hindered cross-border marketing of real estate; it has nevertheless hindered cross-country practice and encouraged non-reciprocity of professional registration between the countries. Notwithstanding, the differences in the courses offered across countries have earlier been attributed to the fact that real estate developed differently in various nations (Mooya, 2007).

In the case of Nigeria and Ghana, the differences only suggest the lack of standardisation of real estate education curricula in Anglophone West Africa. The new desire for regional integration and increase in globalisation has created an urgent need for improvements to meet current global market trends in real estate transactions. Shelton and Howard (2003) and Kampamba et al. (2015) had mentioned that the purpose of analysing real estate curricula is to provide support for educators who are developing their curricula, especially in emerging economies. What is important therefore is the role that such curriculum plays in the production of professionals who meet the standard of what the society demands. Literature on real estate education has shown that there are missing gaps in curriculum development across nations needing improvement (Small & Karantonis, 2001; Crews, 2004; Ayofe, Ajetola, & Oyewole 2009; Black et al., 1996).

Ensuring standardisation in real estate education curriculum is one of the primary concerns of this chapter. A number of options have been shown in literature and table 18.7 presents some notable recommendations from literature that can improve real estate curriculum to meet global standards.

The review has shown notable recommendations for ensuing standardisation. They include a strong real estate curriculum capable of enhancing students' competencies in the marketplace, a curriculum that focuses more on finance, valuations and economics; curriculum that incorporates innovative and flexible modes of learning; curriculum that promote adequate 'technical' and 'soft skills' that can equip students with a foundation for a career in the real estate industry; and curricula of real estate that reflect the real-world market context in which they are delivered.

10. Conclusion

The review has shown that real estate courses and research are composed of interdisciplinary aspects such as economics, law, regional planning, engineering and architecture. The study

TABLE 18.6 Comparison of Courses Offered in Real Estate Polytechnic Programme

S/No.	*Courses offered*	*Ken Saro-Wiwa poly, Nigeria*	*Kumasi poly, Ghana*
1	**Advance Valuation**	√	X
2	**African studies**	X	√
3	**Arbitration & Award**	√	X
4	**Building Construction I**	√	√
5	**Building Economics**	√	√
6	**Building Services and Maintenance**	√	X
7	**Citizenship Education**	√	X
8	**Communication in English**	√	√
9	**Compulsory Acquisition &Compensation Law**	√	X
10	**Computer Application to Real Estate**	√	X
11	**Computer Literacy/Introduction to Computer**	√	√
12	**Elements of Central & Local Government**	X	√
13	**Entrepreneurship**	√	√
14	**Environment & Estate Services**	√	X
15	**Environmental Impact Assessment**	√	X
16	**Environmental Valuation**	√	X
17	**Estate Accounting/Principles Accounting**	√	√
18	**Estate Graphics**	√	X
19	**Estate Office Administration**	√	X
20	**Estate Services**	√	X
21	**Facilities Management**	√	X
22	**Introduction to Estate Management**	√	X
23	**Introduction to Property Development**	√	X
24	**Introduction to Property Management**	√	X
25	**Introduction to Psychology**	√	X
26	**Introduction to Sociology**	√	X
27	**Introduction to Statistics**	√	X
28	**Introduction to Valuation**	√	√
29	**Land Economics**	√	√
30	**Property Law**	√	√
31	**Land Use Planning & Administration**	X	√
32	**Law of Contract and Torts**	√	√
33	**Law Relating to Estate Management**	X	√
34	**Mathematics**	√	√
35	**Mineral Valuation**	√	X
36	**Principles of Law**	√	√
37	**Physical and Health Education**	√	X
38	**Plant &Machinery Valuation**	√	X
39	**Practical Training in Industry**	X	√
40	**Principles & Practice of valuation**	√	√
41	**Principles of land Surveying**	√	√
42	**Principles of Accounting**	√	√
43	**Principles of Economic**	√	√
44	**Property Taxation**	√	X
45	**Principles of Property Rating**	√	√
46	**Professional Practice**	√	X
47	**Professional/Technical Report Writing**	√	√

(*Continued*)

TABLE 18.6 (Continued)

S/No.	*Courses offered*	*Ken Saro-Wiwa poly, Nigeria*	*Kumasi poly, Ghana*
48	**Project Work**	√	√
49	**Property Development**	√	X
50	**Property Management**	√	√
51	**Real Estate Appraisal**	√	X
52	**Research Methods**	√	√
53	**Rural Land Use Economics**	√	X
54	**Technical Drawing**	√	X
55	**Urban Land Use Economics**	√	X
56	**Urban Principles and Techniques**	√	X

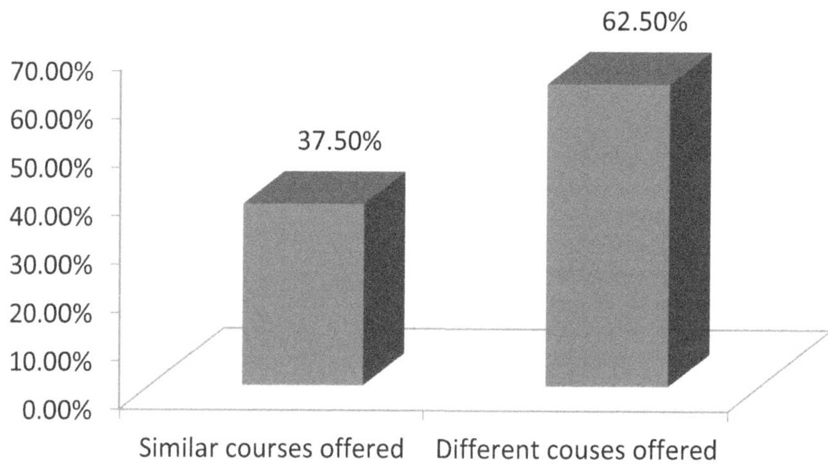

FIGURE 18.2 Comparison of University Course Curriculum in Nigeria and Ghana

revealed that there is a lack of standardisation of the present curricula in real estate education in Anglophone West African. It is believed that real estate education must move in response to the forces which are affecting the profession, and which have a significant and increasing impact in future years (Schulte, 2002). Judging from the findings, it can be inferred that a strong real estate curriculum capable of enhancing students' competencies in the marketplace is what will ensure standardisation. Such a curriculum will focus more on finance, valuations and economics incorporating innovative and flexible modes of learning with a mix of adequate 'technical' and 'soft skills' and will equip students with a foundation for a career in the real estate industry. This curricula must reflect real-world market realities in its rendition of real estate courses. This calls for collaboration between the education and professional institutions to ensure curriculum standardisation in order to maximise the benefit of real estate education in West Africa.

TABLE 18.7 List of Published Real Estate Education Research

S/No.	*Author (s)*	*Year*	*Specifics*	*Recommendations*
1.	Black, R.T. Carn, N.G. Diaz, J. & Rabianski, J. S.	1996	The role of the American Real Estate Society in Defining and Promulgating the Study of Real Property	Creation of a task force to serve as a repository of existing information and to propagate real property education
2.	Small, G. R., &Karantonis, A. C	2001	Property education, training and phenomenalism	Real property courses should incorporate both the classroom education as well as practical training, for the university students to be technically sound and be well equipped to respond to challenges of change.
3.	Weeks, H.S. & Finch, J.H.	2003	An analysis of real estate curriculum requirements at AACSB International-accredited institutions	A strong real estate curriculum that is recognised as enhancing students' ability to compete and succeed in the marketplace is a necessity.
4.	Crews, G. L.	2004	Real estate education on the run: The classroom comes to town	Incorporation of innovative and flexible modes of learning suitable for a nation-wide academic curriculum
5.	Chikafalimani, S.H.P. & Cloete, C.E.	2007	A critical assessment of postgraduate real estate education in South Africa: Importance and research approach	Real estate professionals also need to have knowledge of finance, as well as, the influence that 'social, cultural, legal and political' issues have on real estate.
6.	Blake, A., Andrea, C. & Susilawati, C.	2009	An evaluation of how well undergraduate property students are prepared for commencing their careers	University syllabus should provide graduates with adequate 'technical' and 'soft skills' to equip them to lay foundation for a career in the property industry.
7.	Jay, G. M.	2011	The views of the property industry on the curriculum of the undergraduate property studies degree in the school of construction economics and management at the University of the Witwatersrand	Real estate curriculum should focus more on finance, valuations and economics and less on the quantity surveying and construction type courses, which the current curriculum contains.

(*Continued*)

TABLE 18.7 (Continued)

S/No.	*Author (s)*	*Year*	*Specifics*	*Recommendations*
8.	Adnan, Y. M., Daud, M. N., Alias, A. and Razal, M. N.	2012	Importance of Soft Skills for Graduates in the Real Estate Programmes in Malaysia	Emphasise the importance of key soft skills in the real estate area so as to enhance curriculum development
9.	Ngwoke, M. D. Adedayo, A. M. Olutope, I. V.	2017	Relevance of Entrepreneurship Education in Real Estate Practice for Sustainable National Growth.	Advocated for further training of real estate graduates on business ideas that will assist them in actualising their dreams of becoming employers of labour.
10.	Oloyede, S.A., Iroham, C.O., Ajibola, M.O., & Ayedun C.A.	2017	Real Estate Education in Nigeria: The Need for New Paradigms	Estate management programme should include adventure education courses geared more towards the promotion of soft skills in students.
11.	Mirembe, R. and Viruly F.	2018	Real Estate Knowledge and the Development of a Real Estate Curricula for African Universities	The curricula of real estate should attempt to reflect the market context in which they are delivered.

References

Adnan, Y. M., Daud, M. N., Alias, A., & Razali, M. N. (2017). Importance of soft skills for graduates in the real estate programmes in Malaysia. *Journal of Surveying, Construction and Property, 3*(2), 1–3.

Ayofe, A. N., Ajetola, A. R., & Oyewole, A. S. (2009). Assessment of existing gap between industrial IT skill requirements and computer science curriculum in tertiary institutions. *The Pacific Journal of Science and Technology, 10*(2), 326–336.

Black, R. T., Carn, N. G., Diaz, J., & Rabianski, J. S. (1996). The role of the American real estate society in defining and promulgating the study of real property. *The Journal of Real Estate Research, 12*(2), 183–193.

Blake, A., & Susilawati, C. (2009). An evaluation of how well undergraduate property students are prepared for commencing their careers. *Pacific Rim Property Research Journal, 15*(2), 204–224.

Boakye-Agyeman, N. A. (2006). Polytechnic education in Ghana: The case of the HND estate management programme. Promoting Land Administration and Good Governance, 5th FIG Regional Conference Accra, Ghana, March 8–11.

Boyd, D., Amidu, A. R., & Smith, M. (2013). *Practice-based body of real estate knowledge.* RICS Research. Available at: www.rics.org/us/knowledge/research/research-reports / practice-based-body-of-real-estate-knowledge/.

Chikafalimani, S. H. P. (2010). A critical assessment of postgraduate real estate education in the republic of South Africa. Unpublished doctoral thesis. University of Pretoria, Pretoria.

Chikafalimani, S. H. P., & Cloete, C. E. (2007, June). A critical assessment of postgraduate real estate education in South Africa: Importance and research approach. In *Association of Schools of Construction of Southern Africa-2nd Built Environment Conference* (pp. 17–19). Port Elizabeth, South Africa.

Crews, G. L. (2004). Real estate education on the run: The classroom comes to town. 10th Pacific Rim Real Estate Society (PRRES) Conference Bangkok, Thailand, pp. 25–28.

Dasso, J., & Woodward, L. (1980). Real estate education: Past, present and future- the search for a discipline. *Areuea Journal, 8*, 404–416.

Decree No. 24 of 1975, now CAP E 13, Laws of the Federation of Nigeria LFN, 2007.

Department of Estate Management Handbook (2018). Rivers State University (RSU).

Ghyoot, V. (2002). Real estate education in Africa. In *Schulte Karl-Werner edited "real estate education throughout the world: Past present, & future* (pp. 463–491). Berlin: Springer Science & Business Media.

Information for Programme Re-Accreditation Application of BSc (2016). *Land economy, Kwame Nkrumah university of science and technology*. Kumasi – Ghana: Kwame Nkrumah University of Science and Technology, Kumasi.

Jay, G. M. (2011). The views of the property industry on the curriculum of the undergraduate property studies degree in the school of construction economics and management at the university of the Witwatersrand. Unpublished Master's Dissertation, University of Witwatersrand, Johannesburg.

Kampamba, J., Nkwae, B., & Tembo, E. (2015). A comparative analysis of real estate education curricula in Botswana. *Mediterranean Journal of Social Sciences, 6*(5), 105.

Millington A. F. (2003). *Professional issues for property people. Warour street.* London: Estate Gazette.

Ministry of Education (1993). *White paper on the reforms to the tertiary education system, Accra.* Accra: Ministry of Education.

Mirembe, R., & Viruly, F. (2018). Real estate knowledge and real estate curricula in African universities. *Journal of African Real Estate Research, 3*(2), 9–17. DOI: 10.15641/jarer.v3i2.692.

Mooya, M. M. (2007). The impact of property education on property research in South Africa: A review. *RICS Research Paper Series*, 7(6), 9–32.

National Board for Technical Education NBTE (2002). *National diploma in estate management curriculum and course specifications.* Kaduna: NBTE.

Ngwoke, M. D. A., Adedayo, M. L., & Olutope, I. V. (2017). Relevance of entrepreneurship education in real estate practice for sustainable national growth European. *Journal of Educational Sciences, EJES Edition, 4*(2), 121–135.

Nigerian Institution of Estate Surveyors and Valuers Code of Professional Ethics and Practice.

Nsiah-Gyabaah, K. (2005). Polytechnic education in Ghana: The past, present and the future. A Paper presented at the Kick-off Conference: NPT/UCC Project on Building Management and Leadership Capacity in Polytechnics at the University of Cape Coast. 20–22 May 2005.

Oloyede, S. A., Iroham, C. O., Ajibola, M. O., & Ayedun, C. A. (2017). Real estate education in Nigeria: The need for new paradigms. Proceedings of EDULEARN17 Conference 3–5 July 2017, Barcelona, Spain.

Poon, J., Hoxley, M., & Fuchs, W. (2011). Real estate education: An investigation of multiple stakeholders. *Property Management, 29*(5), 468–487.

Sawyer, H. M. (1972). Presidential speech at thc inauguration of the Ghana Institution of Surveyors. *The Journal of the Ghana Institution of Surveyors, II*(1), 19–21.

Shelton, H., & Howard, J. (2003). An analysis of real estate curriculum requirements at AACSB international-accredited institutions. *Journal of Real Estate Practice and Education, 6*(2), 257–268.

Schulte, K. W. (2002). *Real estate education throughout the world: Past present, &future.* London: Kluwer Academic Publisher.

Schulte, K. W., Schulte-Daxbök, G., Holzmann, C., & Wiffler, M. (2005). Internationalisation of real estate education. European Real Estate Society Conference: June 2005, Dublin, Ireland.

Small, G. R., & Karantonis, A. C. (2001). Property education, training and phenomenalism. In *Pacific rim real estate conference* (pp. 1–8). Pacific Rim Real Estate Society. Available at: https://opus.lib.uts.edu.au/bitstream/10453/7357/1/2005001223.pdf.

Thorncroft, M. (1965). *Principles of estate management*. London: Estate Gazettes Ltd.
Umeh, J. A. (1986). The faculty of environmental studies. In E. Obiechina et al. (eds.) *University of Nigeria 1960–1985: An experiment in higher education*. Nsuka: University of Nigeria.
Weeks, S. H., & Finch, H. J. (2003). An analysis of real estate curriculum requirements at AACSB international accredited institutions. *Journal of Estate Practice and Education*, *6*(2), 257–268.
Yu, S. (2001). New paradigms in real estate education. *Pacific Rim Property Research Journal*, 7(2), 79–88.

INDEX

Note: Page numbers in *italics* refer to figures; those in **bold** refer to tables.

Abidoye, R. B. 115, 116
actuators 189
Adams, R. H., Jr. 12
Adebiyi, S. O. 115
Adeboye, A. 113
Adesope, A. A. 169
Adewunmi, Y. 113
Adjekophori, B. 115
administration 193
Africa; *see also* housing building guidelines in Africa: colonial occupation of 22–23; countries covered by GRETI 6; market maturity in, real estate industry 213–214; neolithic settlements in 20; suburban neighbourhood designs in 23
Africa Continental Free Trade Agreement (AfCFTA) 45
The African City: A History (Freund) 20
African Continental Free Trade 11
Africanness, cultural statement of 20
African property markets: global trends in 4–15; introduction to 1–3; residential and commercial, West Africa 37–46
African Real Estate Society 106
African Staff Housing Fund 171
Agence Française de Développement (AFD) 12
Ahmad, M. 79
Ajayi, O. 115
Akaabre, P. B. 116
Akinlabi, A. J. 114–115
Akinsomi, O. 178, 181
Albert, J. D. 106
Alkali, M. 116
Allen, P. J. 188
Amadi-Echendu, A. P. 203, 204, 206, 207–208
Ampofo, J. A. 117
Anderson, R. I. 180
Anim-Odame, W. 113
apartheid 24; housing shortages under 122–124
Appolonia (Accra) 27
The Appraisal Journal (TAJ) 106
Archer, W. R. 158
Asabere, P. 111
Attride-Stirling, J. 206
Awuah, K. G. B. 24, 25
Ayentimi, D. T. 204, 209

Babawale, G. K. 113
Bah, El-hadj M. 37
Balde, Y. 12
Batubara, F. R. 204, 209
Baum, A. 202–203, 204, 205, 209
behavior inference systems 193
Bell, J. 206
Benefield, J. D. 178, 180, 186
Berges, S. 168
Black, R. T. 230
blockchain technology: defined 201; digital technology in transaction process and 204–205
BMS *see* Building Management System (BMS) in Kenya
Boamah, N. A. 112
Boles, J. R. 203

Bondinuba, F. K. 115
Boore, C. 77, 84
Borras, S. M., Jr. 75
Boyd, D. 188, 229
Breaking New Ground (BNG) 138, 139
Bretton Woods Institutions 24
British Town and Country Planning Act 25
Broll 6
Brown, G. R. 5
Brown, R. G. 188
Brownfields 128
Bruce, J. 76, 77, 78, 79, 80
Building Automation System (BAS) 188; *see also* Building Management System (BMS) in Kenya
Building Management System (BMS) in Kenya 188–199; applications of, commercial property management 190–191, *191*; BMS defined 188; challenges/barriers to 193–194, 196, 197, *197*, 198; commercial building implementation of 195, 196, *197*; components of 189, 190; concepts/technologies, awareness and adoption levels of 194–195, *196*; data analysis/results 194–198; described 189, *190*; fire detection/safety systems and 191; HVAC and 192; introduction to 188–189; lighting controls and 192; parking control and 193; recommendations for 198, 199; research design/methodology 194; security/access controls and 191–192; voice/videocon services and 193; widespread adoption strategies for 198, 198, *199*
Burgess, J. 204, 209
Burns, T. 78

Calthorpe, P. 130
Campbell, K. 32, 33
Carn, N. G. 216
CBRE 6
Central Bank of Nigeria (CBN) 171
Chan, A. P. 115, 116
Chigbu, U. E. 49
Cho, H. 181
Chukwu, A. C. 160
Ciesielska, M. 193, 196, 198
Clauretie, T. 106
Clinton Foundation 127
Cloete, C. E. 111, 112, 207
colonialism and apartheid, housing shortages under 122–124
colonies 122
Comité Français pour la Solidarité Internationale (CFSI) 12
communal land tenure classification 50
Community Land Act, 2016 50, 101
Community Land Act, 2016a 95
Community Residential Units (CRU) 131
competing interests, land tenure 49
complementary interests, land tenure 49
contemporary urbanisation patterns 26–27
controllers 189
County Government Act, 2012 97
COVID-19 pandemic: impact on Gauteng Housing Demand Model 135, 150–151, 151, 152; impact on Nigeria real estate markets 166; sub-Saharan Africa 14–15
Crews, G. L. 230
Crown Lands Ordinance 1902 75
customary land tenure system 51–52; advantages of 52; characteristics of 51–52; limitations of 52

Daneshvary, N. 106
Das, S. 113
Dasso, J. 229–230
data transparency, mature markets and 5
de Boer, E. 209
decentralization: of authority over land administration 82; concepts of 78; dimensions of 78–79; facets of 77; fiscal 79; institutional 79; land administration 80–82; levels of 77; need for 77–78; political 79
decentralized land administration, Kenya 75–89; authority over land and 82; cautious approach to 79–80; concepts of 78; described 80–82; devolution and 83; dimensions of 78–79; functions and institutions 84–85; funding 85–86; future for 88–89; human capacity 85; introduction/background to 75–76; land information system 86; legal framework 86–87; need for 77–78; relationship model 87, *87–88*, 88; review of 83–87
deconcentration 78, 82
deforestation 59
delegation 78
delocalization 78
demand models described 140–141
demographics, Sub-Saharan Africa 9–10
Department of Human Settlements (DHS) 138
de-risking inclusive and affordable housing 126–129
De Soto, H. 29, 30, 114, 165, 169, 203
devolution 78, 82, 83
Dewar, D. 31
Diaz, J., III 106
Digital Ledger Technology (DLT) 203

digital technology in transaction process 201–210; advantages of implementation of 208; African context literature 205–206; barriers to implementation of 208–209; disadvantages of implementation of 208; further research for 210; impact of, on South African property transaction process inefficiencies 207–208; introduction to 201–202; research methodology 206; South African property transaction process inefficiencies and 206–207; technological development literature 202–203; through blockchain technology literature 204–205
Dipholo, K. 203, 206
discount-benefit scheme housing 138, 139
Dutch East India Company 22, 23

Ease of Doing Business Index 63, 68–69; Nigeria 70–72
East African (lands) Order-in-Council 75
Eco city (Lagos) 27
eco-friendly and smart housing developments 128–129
eco-friendly housing 126
economic viability, land tenure systems and 58–59
economy, land tenure and 68–69
Edwards, W. K. 193
Elkes, L. 216
Embaye, W. T. 117
enclavements 24
Energy Act, 2019a 94
Energy Act of 2019 98–99
Energy and Petroleum Regulatory Authority (EPRA) 99
Energy and Petroleum Tribunal (EPT) 99
Environmental Impact Assessment (EIA) 95
Environmental Management and Co-ordination Act (EMCA), 1999 94, 95–96
environmental viability, land tenure systems and 59–60
Environment and Land Court Act, 2011 97
Epley, D. 215
Estate Intel 160
Estate Surveyors and Valuers Registration Board of Nigeria (ESVARBON) 233, 240
Evans, K. 110
exclusion patterns 20
Ezimuo, P. N. 170
Ezinwanne Udechukwu, C. 112

fantasy cities 27–28
Federal Mortgage Bank of Nigeria (FMBN) 171
Federal Mortgage Finance Limited (FMFL) 171
financial barriers, BMS 193
financing infrastructure through incentive-based system 127
Finch, H. J. 216
fire detection/safety systems, BMS 191
fiscal decentralization 79
Food and Agriculture Organization (FAO) 57, 80
Ford, D. 216
foreign direct investments 10–11; Sub-Saharan Africa 10–11
Forest Management and Conservation Act, 2016b 95
Forest Management and Conservation Act 2016 100–101
formal and legal settlements 23
4Ps *see* Public Private People Partnership Inclusive Housing Delivery Model (4Ps)
Fourth Industrial Revolution (4IR) 202
Franco, C. J. 75
freehold tenure 53; advantages of 54; limitations to 54
Freund, B. 20, 23, 31
FTSE EPRA NAREIT global indices 176, 177

Garden Cities 23
gated communities 110, 123
Gauteng Department of Human Settlements (GDHS) 135
Gauteng Housing Demand Model 135–155; background 135–141; business as usual scenario 147, *147–148*, 148; COVID-19 pandemic impact on 135, 150–151, 152; dashboard *142*; demand models described 140–141; dimensions for backlogs 144–145; external factors affecting future projections 141, 143; Gauteng province described 136; high-level overview of 145, *146*; housing described 136, 138, 139, 140; increased private sector involvement scenario 148–149, *149–150*, 150; introduction to 135; municipalities and wards *137*; overview of 140, 141, 143–147; pseudo decision algorithm 145, *146*, 147; recommendations from 152–154
Gauteng province 136; housing in 136, 138, 139, 140
Gauteng Provincial Economic Review and Outlook (PERO) 147
Gavu, E. K. 117
Gbadegesin, J. T. 114
Gbonegun, V, 173
Geogievski, B. 162

Geyer, H. S. 114
Geyer, H. S., Jr. 114
Ghana Real Estate Developers Association (GREDA) 41
Ghana real estate markets: commercial/office space 42–43, *43*; dynamics of 40; overview of 40–41; rental values and yields 43; residential 41–42; residential accommodation types in *42*; retail *43*, 44; transparency in *40*
Ghebru, H. 69
Gibler, K. M. 106
Gilbert, A. 29–30
Gleave, F. G. 232
Global Real Estate Transparency Index (GRETI) 5–6, 39, 163, 202
global trends in African real estate 4–15; COVID-19 pandemic and 14–15; demographics and 9–10; foreign direct investments and 10–11; introduction to 4–6; macroeconomic regimes and 6–7; migration patterns and 7–8; mortgage market and 13–14; property technology and 12–13; remittances and 11–12; urbanisation and 8–9
Government Lands Act (G.L.A.) 76
Graglia, J. M. 203, 207, 208
Graham, N. 127
Grant, R. 110, 111
Green Building Technologies (GBT) 128
Greenfields, building inclusive housing in 127–128, 130
GRETI *see* Global Real Estate Transparency Index (GRETI)
Grinter, R. E. 193
Group Areas Act 41, 1950 123
Guardian, The (newspaper) 172, 173
Gulyani, S. 29, 111–112
Gumede, N. 203, 206
Gunter, A. 113–114
Gupta, S. 11
Gyamfi-Yeboah, F. 112
Gyourko, J. 180

Hardin, W. G., III 106
heating, ventilation and air conditioning (HVAC) 188, 192, 194
Highest and Best Use (HBU) concept 127
high to medium value urban development and real estate investments 27–28
Hillhorst, T. 83
homelands 122
Housing Act 32, 1920 123
housing building guidelines in Africa 122–131; applying 4Ps 126–127; building inclusive housing in Greenfields 127–128; colonialism and apartheid, housing shortages under 122–124, *124*, *125*; de-risking inclusive and affordable housing 126–129; eco-friendly and smart housing developments 128–129; financing infrastructure through incentive based system 127; future guidelines for 129–131; inclusive housing 124, 126; introduction to 122; market implications 131; pepper potted topologies in inclusive housing developments 128; policy implications 131; private sector-oriented asset/property management strategy 129
housing opportunity 138
Howard, J. 229, 244
Huchzermeyer, M. 29
HVAC *see* heating, ventilation and air conditioning (HVAC)

Ikekpeazu, O. F. 116
incentive based financing infrastructure 126
inclusion patterns 20
inclusive housing 124, 126; eco friendly and smart 128–129; in Greenfields, building 126, 127–128; guidelines for future 129–131; pepper potted topologies in 128
informal and illegal settlements 23, 25, 29
informal sector owner-developers 25
innovative digital technology (IDT) 201, 202
institutional decentralization 79
International Housing Solutions 127
International Monetary Fund (IMF) 24, 29
Internet of Things (IoT) 129, 192
interoperability 193
investor outlook, West African real estate markets 39
Irumba, R. 114

Jankovic, L. 188
Jones Lang LaSalle (JJL) 5, 6; Global Real Estate Transparency Index 5–6, 39, 163, 202
Journal of Real Estate Finance and Economics, The (JREFE) 106
Journal of Real Estate Literature 106
Journal of Real Estate Research (JRER) 106
Journal of the American Real Estate and Urban Economics Association, The 106
Journal of Urban Economics, The (JUE) 106
Jupp, B. 124, 126

Kajimo-Shakantu, K. 110
Kamati, K. 116

Kampamba, J. 213, 229, 244
Karam, A. 29, 106
Karantonis, A. C. 228, 230
Kasimbazi, E. 49
Kaulihowa, T. 116
Kauzya, J. 78, 79
Keeton, R. 28, 32
Kenya; *see also* Building Management System (BMS) in Kenya; decentralized land administration, Kenya; sustainable development in Kenya: community land category 50; customary law 51–52; land tenure systems in 48, 51–54; National Land Policy 49, 57; private land category 50; public land category in 50; statutory law 51, 53–54
Kenya Human Rights Commission 51
Kenyan Constitution 57
Kenyan National Land Policy 49, 57
Kibunyi, D. 115
King, R. 124
Knight, J. R. 110
Knight Frank 6
Knox, D. 76, 77, 78, 79, 80
Kolvart, M. 203, 204, 208
Kozma, M. 204, 207
Kshetri, N. 204, 208
Kuma, S. S. 114
Kumar, A. 204, 209

laissez-faire developments 24–25
Land Act, 2012a 94
Land Act 2012 101
land administration decentralization 80–82
Land and Native Rights Act 1916 66
Land Economics (LE) 106
land grab 23
Land Proclamation Act 1900 66
land registration, described 64
Land Registration Act 57; 2012 101; 2012b 94–95
land rights, protection/security of 57–58
Lands and Native Rights Act 1916 66
land security, defined 49
Land Tenure Law of Northern Nigeria 1962 67
land tenure systems 48–60; *see also* Nigerian land tenure impacts; sustainable land management (SLM); access to land and 56–57; classification of 49–50; concept of 48–49; definitions of 48–49; described 48; economic viability and 58–59; environmental viability and 59–60; intersecting interests of 49; introduction to 48; in Kenya 48, 51–54; protection/security of land rights and 57–58; social acceptability and 60; sustainable land management and 56–60; termed 63
Land Use Act 1978 67–68
land use rights, described 64
leap-frog developments 27
leasehold tenure 53, 54; advantages of 54; limitations to 54
legal/financial considerations, African property markets: decision making impacts; global market forces and; market outcomes and; transparency and
Leurent, H. 209
Levin, S. 204, 207, 208, 209
Li, F. 193, 196, 198
lighting controls, BMS 192
Lin, Y. C. 177, 178, 180, 184, 185, 186
Ling, D. C. 158
Liu, N. 206
low value real estate developments and urban growth management 28

macroeconomic regimes 6–7
Manda, M. A. Z. 111
Mansfield, E. 205
Marais, L. 114
marginal value urban developments 29–30
market competitiveness 159
Massyn, M. W. 126
Matongela, A. M. 114, 116
mature markets, data transparency and 5
Matysiak, G. A. 5
MCORE 161, 162
Megbolugbe, I. F. 108
Mehdi, N. 208
Mellon, C. 203, 207, 208
Microsoft 11
migration patterns 7–8; defined 7; relevancy to real estate markets 7–8
Miller, J. 123
Mirembe, R. 214
modern economy, land tenure and 68–69
Mooya, M. M. 112, 207, 229
Mortgage Institutions Act, 1989 171
mortgage market, sub-Saharan Africa 13–14
Moss, A. 180
Mugabe, J. 49
Mugenda, A. G. 194
Mugenda, O. M. 194
Mvuyana, B. Y. C. 126–127
Mwakubo, S. 59–60
Mwangi, W. 80, 81

Myeni, S. L. 126–127
Mystery of Capital. The (De Soto) 165

NACI 205
Nadiri, M. I. 205
National Bureau of Statistics (NBS) 172
National Environment Management Authority (NEMA) 95
National Environment Tribunal (NET) 96
National Housing Fund Act, 1992 171
National Housing Fund (NHF) program 171
Native Land Act 27, 1913 122–124
Native Lands Acquisition Act 1917 66
Native Land Trust Ordinance 76
Natives Urban Areas Act 35, 1923 123
Ndinda, C. 111
Neighbourhood unit 23
Nelling, E. 180
Newell, G. 180–181
New Town Movement 23
Nigerian Building Society (NBS) 171
Nigerian Institution of Estate Surveyors and Valuers (NIESV) 173, 233, 240
Nigerian land tenure impacts 63–73; agency staffing issues and 71; Ease of Doing Business Index 70–72; evolution history of 64–68; introduction to 63–64; Land Use Act provisions and 70–71; market environment constraints and 70; modern economy and 68–69; property market operations and 69–72; recommendations for 72–73; registration/fee costs and 71–72; transparency and 72
Nigerian Mortgage Refinance Company (NMRC) 171
Nigeria real estate markets 158–166; *see also* real estate finance challenges in Nigeria; apartment hotels (short lets) opportunities 163; COVID-19 pandemic impact on 166; co-work office space opportunities 163; dynamics of 40; financing challenges 164–165; foreign exchange challenges 165; green property markets 161; industrial sector 45; infrastructure challenges 164; institutional investor opportunities 164; introduction to 158–159; joint venture opportunities 163; office 44; opportunities in 163–164; overview of 44; property market data challenges 165–166; property market review 159–163; residential 44; residential real estate/shops opportunities 164; retail 45; terrorist activists/insurgency challenges 166; title registration challenges 165; transparency in *40*
Niger Land Transfer Act 66
Nijhuis, S. 28, 32
Nkechi, O. T. 115–116
Northcourt Real Estate 161
Northern Land Tenure Law 1962 66
novel corona pandemic *see* COVID-19 pandemic
Nowiński, W. 204, 207
Ntuli, M. 178, 181
Nubi, T. G. 112
Nyambe, C. 116
Nyametso, J. K. 113
Nyasulu, E. C. 111
Nyika, D. 80, 81

Obeng-Odoom, F. 112
Obunde, O. P. 56
Ogendegbe, P. S. 169
Ogendo, O. 51
Ogolla, B. 49
Ogutu, C. 189
Okumo, A. 69
Oloyede, S. A. 230
Oluwamotemi, D. K. 67
Onibokun, G. A. 108
O'Sullivan, A. 20
Otegbulu, A. C. 161
overlapping interests, land tenure 49
overriding interests, land tenure 49
Owusu-Ansah, A. 115, 117
Owusu-Manu, D. G. 116

parking control, BMS 193
Pellissier, R. 203, 204, 206, 207–208
pepper potting 126; in inclusive housing developments 128
Physical and Land Use Planning Act, 2019 96–97
Physical and Land Use Planning Act, 2019b 94
Physical and Land Use Plans, categorized 96
political decentralization 79
Poon, J. 230
Population Registration Act, 1950 123
private land tenure classification 49–50
private oriented asset/property management tools 126
private sector investment 24–25
private sector-oriented asset/property management strategy 129
property market, defined 64
property technology (PropTech): defined 12; purpose of 12–13; sub-Saharan Africa and 13
property transaction process 201
PropTech sector 202–203

Protection of Personal Information Act 4, 2013 209
Public Health Act Cap 242 97–98
Public Health Act Cap 242, 2017a 94
Public Lands Acquisition Act 1917 66
Public Lands Acquisition Act 1947 67
public land tenure classification 50
Public Private People Partnership Inclusive Housing Delivery Model (4Ps) 126; applying 126–127
PWC 27

Quayson, A. 111

Rabianski, J. 216
Rama, U. 171
real estate: defined 23; global level, defined 5; initial record of term 23; investible values of 5; market performance in some African countries 38
Real Estate Economics (REE) 106
real estate education, West Africa 228–248; career opportunities 231–232; curricula comparison 241–244, 242–243, *243*; entry requirements 235; Ghanaians education and training in real estate 238; graduate programmes in 236; introduction to 228–229; literature review 229–230; Nigeria and Ghana curricula comparison 243–244, *246*; Nigerian education/training in 232–237; polytechnic, description and core curriculum 235–236, 237, 239–240, 245–246; profession of real estate practitioners 240; programme philosophy 230–231; published real estate education research 247–248; research method used 232; standardisation of curriculum and 244; undergraduate programmes in 238, 239; university under-graduate programme 233, 234, 235
real estate education south of Sahara 213–225; categorisation of 219, *220*; introduction to 213–214; literature review 214–216, 215; programmes offered for 218–220, 221–222, 223–224; recommendations for 224–225; research methodology 216–218; summary of existing studies on 217
real estate finance challenges in Nigeria 164–165, 168–175; capital market awareness and 174; citizen awareness and 174; commercial banks and 170; equity finance sources and 172; financing sources 170–174; forward sale/off plan 172–173; interest rates charged and 174; introduction to 168–169; literature concerning 169–170; mortgage banks and 171; mortgage structures and 173; real estate funding constraints and 173; real estate investment trusts 173; reasons for 169; title registration difficulties and 174
real estate investments, property development in African cities and: for colonial settlements 22–23; high to medium value 27–28; introduction to 18–19; low value 28; marginal value 29–30; neo-liberal influence on 24–26; states with independence and 23–24
real estate investment trusts (REITs) 173; *see also* South African real estate investment trusts (SA REITs); diversified 178, 179; sector-specific 178, 179
real estate transparency in sub-Saharan Africa *40*
Reconstruction and Development Programme (RDP) 138, 139
Redman, A. L. 106
Reed, R. 160
reliability 193
remittances in sub-Sahara Africa 11–12
Remke, R. 188
residential markets, Ghana 41–42, *42*
Restructuring Capital Grant 131
Restructuring Capital Grant Quantum 131
RICS Valuation – Global Standards (Red Book Global standards) 15
Rio Declaration on Environment and Development 91–92, 94
Ro, S. 180, 186
Rogers, E. 193
Romer, P. M. 205
Roy, A. 29, 30
Royal Institution of Chartered Surveyors (RICS) 15, 230, 240
Rural Electrification and Renewable Energy Corporation (REREC) 99

Sani, K. S. 114
Sanz, M. J. 55
Saull, A. 204, 206, 209
Scheepers, L. 113
"Scramble for Africa" 22
Secondary Mortgage Markets (SMM) 112
security 193
security/access controls, BMS 191–192
sensors 189
SERI 207
Seso Global 206
Sessional Paper No 3 of Land Policy, 2009 76
Shelton, H. 229, 244

Shimelles Tenaw, K. M. 56, 57
silo mixing 124, 126
Simons, R. A. 106
Sims, S. 160
Singh, R. 12
Sky Shelter Fund 173
Small, G. R. 228, 230
Small Medium and Micro Enterprises (SMMEs) 131
smart affordable housing 126
smart contracts 201
smart housing developments 128–129
Smoke, P. 79, 80
social acceptability, land tenure systems and 60
social barriers, BMS 193
Social Housing Regulatory Authority (SHRA) 131
Social Security and National Insurance Trust (SSNIT) 41
Solow, R. M. 205
South African Cities. A Manifesto for Change (Dewar and Uytenbogaardt) 31
South African real estate investment trusts (SA REITs) 176–186; active domiciled 182; diversification benefits 184–186, 185; growth rate of *177*, *178*; introduction to 176–178; literature review of 180–181; performance *183*; profile of leading 179; profile of sector-specific 179; rank of 177; regulation of 176; research data and methodology 181–182, *183*; risk-adjusted returns 184; understanding 178–180
Soyendo, V. 172
Spielman, A. 208
State Housing Company (SHC) 41
Stats SA 141
statutory land tenure systems 51, 53–54; freehold tenure 53; leasehold tenure 53, 54
Strategic Environmental Assessments (SEA) 95
structural barriers, BMS 193
sub-Saharan Africa (SSA): demographics 9–10; foreign direct investments 10–11; middle-income class 11; real estate markets 4; urbanisation 8–9, 18
sub-Saharan Africa (SSA) real estate research, evolution of 105–117; analysis method described 106, 107; chronological themes analysis 108, 109–117; geographic extent of research 108, 109; introduction to 105; research agenda in 105–106, 107
Supervised Industrial Work Experience (S.I.W.E.) 233
supply and demand models 140
Sustainable Development Goals (SDGs) 32
sustainable development in Kenya 91–102; agenda 91–92; Community Land Act, 2016 101; constitutional provisions 92–94; Energy Act of 2019 98–99; Environmental Management and Co-ordination Act, 1999 95–96; Forest Management and Conservation Act 2016 100–101; introduction to 91; Land Act 2012 101; Land Registration Act 2012 101; Physical and Land Use Planning Act, 2019 96–97; Public Health Act Cap 242 97–98; real estate legislation affecting 94–101; sectoral policies affecting 101–102; Water Act 2016 99–100
sustainable land management (SLM) 55–60; *see also* land tenure systems; definitions of 55; land tenure systems and 56–60; objective of 55; principles and criteria for 55
Swan, M. 204
Syagga, P. M. 76
Systematic Land Titling and Registering project 69
systems intelligence 193

Talukdar, D. 29, 111–112
Tatu city (Nairobi) 27
TechHubs 11
Tema Development Company (TDC) 41
tenure security concept 49
territories 122
Teye, J. K. 115
Thompson, L. A. 123
Tiffen, M. 60
Tomlinson, M. R. 111
Toulmin, C. 203
Town Planning Review (journal) 105
Treaty of Cessation 66
Trust Land Act Cap 288 82

Ubogu, R. E. 108
Ubosi Eleh & Co 162
UN-Habitat 32, 48, 49
Union Homes Real Estate Investment Trust 173
United East India Company 22
UPDC Real Estate Investment Trust 173
Upton, C. 82
Urban Areas and Cities Act 2011 97
urban growth, property development and 18–33; in ancient neolithic settlements 20, 21–22; colonial settlements and 22–23; contemporary urbanisation patterns 26–27; framework overview 19–20; high to medium value urban development 27–28; impact on real estate

market activities 30–32; independence and 23–24; introduction to 18–19; low value urban development 28; marginal value urban development 29–30; neo-liberal influence on 24–26
urbanisation: global rate of 18; global trends in 8–9; Sub-Saharan Africa 8–9, 18
Urban Settlements Development Grant (USDG) 131
Usman, H. M. 169
Uson Eleh & Co 161–162
Uytenbogaardt, R. 31
Uzodike, U. 111

Vancouver Declaration on Human Settlements, 1976 124
Veit, P. 58
Verbong, G. P. 193
Viruly, F. 214
voice/videocon services, BMS 193

Wang, S. 188
Wanjiru, B. 189
Water Act, 2016c 94
Water Act 2016 99–100
Waterfall City (Johannesburg) 27
Water Resources Authority (WRA) 100
Water Services Regulatory Board (WSRB) 100
Watson, V. 20, 27, 29
Webb, J. R. 106
Weeks, H. S. 106, 216
Wen, H. 180–181
West Africa residential/commercial markets 37–46; in Ghana 40–44; introduction to 37–39; investor outlook in 39; in Nigeria 40, 44–45; performance of 38; transparency in *40*
women land rights 58
Wong, P. K. 205
Woodward, L. 229–230
Woonerf Concept 23
Workforce Housing Fund 127
World Bank 12, 24, 29, 31, 55, 123, 205; Doing Business 2020 174; Ease of Doing Business Index 63, 68–69, 165
World Health Organization (WHO) 14, 129

Xie, J. 188

Yu, S. 229

Zheng, Z. 203, 207
Ziobrowski, A. J. 106, 112, 180, 186

Taylor & Francis eBooks

www.taylorfrancis.com

A single destination for eBooks from Taylor & Francis with increased functionality and an improved user experience to meet the needs of our customers.

90,000+ eBooks of award-winning academic content in Humanities, Social Science, Science, Technology, Engineering, and Medical written by a global network of editors and authors.

TAYLOR & FRANCIS EBOOKS OFFERS:

- A streamlined experience for our library customers
- A single point of discovery for all of our eBook content
- Improved search and discovery of content at both book and chapter level

REQUEST A FREE TRIAL

support@taylorfrancis.com

Routledge — Taylor & Francis Group
CRC Press — Taylor & Francis Group

For Product Safety Concerns and Information please contact our EU representative GPSR@taylorandfrancis.com Taylor & Francis Verlag GmbH, Kaufingerstraße 24, 80331 München, Germany